Cosmology 2020 - The Current State

Edited by Michael L. Smith

Published in London, United Kingdom

IntechOpen

Supporting open minds since 2005

Cosmology 2020 - The Current State
http://dx.doi.org/10.5772/intechopen.87805
Edited by Michael L. Smith

Contributors
Gabriel Vacariu, Mihai Vacariu, Constantin Meis, Michael L. Smith, Ahmet M. Mecit Öztaş, Jan Olof Olof Stenflo, Celia Escamilla-Rivera, Eugene Terry Tatum

First published in London, United Kingdom, 2020 by IntechOpen
IntechOpen is the global imprint of INTECHOPEN LIMITED, registered in England and Wales, registration number: 11086078, 7th floor, 10 Lower Thames Street, London, EC3R 6AF, United Kingdom
Printed in Croatia

British Library Cataloguing-in-Publication Data
A catalogue record for this book is available from the British Library

Additional hard and PDF copies can be obtained from orders@intechopen.com

Cosmology 2020 - The Current State
Edited by Michael L. Smith
p. cm.
Print ISBN 978-1-83968-267-4
Online ISBN 978-1-83968-268-1
eBook (PDF) ISBN 978-1-83968-269-8

For EU product safety concerns:
IN TECH d.o.o., Prolaz Marije Krucifikse Kozulić 3, 51000 Rijeka, Croatia, info@intechopen.com or visit our website at intechopen.com.

Meet the editor

Dr. Michael Smith has decades of experience in computerised data analysis and has co-authored many articles on novel ideas in cosmology and fundamental physics. He recently developed a more general version of the Einstein field equation. He has analysed many variations of Λ cosmology using the FLRW model using supernovae Ia, II and gamma-ray burst data. In addition, he co-authored one of the first articles on polytropic cosmology published by IntechOpen. He is currently interested in the real meanings of the spacetime parameters, K and Ωk, of the FRW model. After many years working as a consultant for both the pharmaceutical and energy industries, he has recently retired to spend some of his time writing about the philosophy of cosmology.

Contents

Preface

Cosmology is necessarily an all-encompassing endeavour. One cannot consider galactic groups without the spacetime in between, galaxies without stars and stars without considering fusion physics. We cover several areas of cosmology with this book, from a philosophy as applied to current cosmology, through dark energy and dark matter via analyses of HII and SNe Ia data, to investigation of the vacuum state. We have allowed all of our authors the freedom of thought and expression so often denied investigators by other means. The result is a collection of fresh new ideas and we hope some of these will interest the reader.

The first and broadest chapter, by Drs Gabriel and Mihai Vacariu, dives into some of the specifics of the philosophical meanings of dark energy and dark matter. They point out that in theory and in reality, one cannot ignore the microscopic world to concentrate solely on the Universe at the galactic level (and greater). They also strongly suggest that allowance for the very large and the very small is not unconditional for all cosmological philosophies. Allowance for both types of science, investigation of the microscopic and gravitational physics, is dependent on the initial conditions of our Universe(s) and the states of our Universe immediately after the big bang. This work is a nice continuation of their earlier thoughts on these topics.

In her chapter about deep learning, Dr Celia Escamilla-Rivera outlines the requirements and techniques used by machine learning methods, sometimes called Artificial Intelligence (AI), to answer some questions of cosmology. She uses the recently preferred Bayesian method for evaluation of some models with the typically large and useful data sets now available to cosmologists. This chapter is a continuation of her recent work in this exciting field. Since computerised applications using AI techniques are rapidly increasing in sophistication, we will hear much more of this pursuit type in the near future.

The Hubble-Lemaitre constant, H_0, is a key value required for understanding the origin, evolution and fate of our Universe. Transformed data are most often used in calculations by those who are now refining the value of H_0. Independent groups utilising gravitational waves and others relying on signals from the CMB and the tip of red giant star branch all estimate a $H_0 \sim 70$ km s^{-1} Mpc^{-1}. This estimate has created 'tension' between two groups employing SNe Ia and HII data who independently calculate a $H_0 \sim 74$. In this chapter, Dr Smith and Dr Öztas describe some common pitfalls and questionable results when log-transformed HII/GEHR and SNe Ia data are used, rather than the actual distance estimates. They also point out that results are more meaningful when low quality HII and GEHR data pairs are discarded. They test six important cosmological models using HII/GEHR data with a much better analytical routine but produce no clear winner.

Dr Jan Stenflo presents a new explanation for the mechanism now forcing our Universe to expand at an accelerating rate. The most common explanation agrees with the suggestion of Einstein that a new energy form is responsible, which is now termed "dark energy". Stenflo rather suggests the Λ term is induced by a global boundary constraint tying that value to the conformal age of our Universe. This solution avoids the cosmic coincidence problem and the problem of the Λ sign change. (Λ supposedly went from being attractive to repulsive several billion years ago.) He illustrates how the cosmological evolution implied by this condition, without introducing free parameters, differs from the ΛCDM standard model. He goes on to predict a current value for Λ within 2σ of that derived from recent CMB observations of the Planck satellite. Using this model, the Universe is calculated to be mildly inflationary throughout the entire radiation-matter dominated epochs. This model also obviates the requirement for a violent GUT-era inflation to explain the large-scale homogeneity and isotropy of our Universe.

Dr Constantin Meis investigates the electromagnetic field ground state in his chapter. He begins by considering a zero-energy vacuum component issuing naturally from Maxwell's equations. In his theory, the vector potential quantization at a single photon level overcomes the vacuum energy singularity in quantum electrodynamics. This singularity leads to the well-known "vacuum catastrophe" in cosmology. He reasonably considers photons as oscillations of the electromagnetic field ground state, which is composed of a real universal electric potential permeating all spacetime, being the origin of the Fulling-Davies-Unruh effect. Fluctuations of the electromagnetic field ground state contribute to the cosmic electromagnetic background and may be the origin of dark energy. He goes further in expressing the gravitational constant through the quantized amplitude of the electromagnetic field ground state, thus connecting electromagnetism with gravity. This development opens new avenues of possibility for further investigation by cosmologists.

Dr E. T. Tatum suggests that what we currently refer to as "cold dark matter" is rather, slow-moving, nearly collision-less interstellar and intergalactic neutral atomic hydrogen in its lowest, 1s, ground state. The density of neutral H in the spacetime vacuum is estimated from the intensity of the signature spectral hyperfine, 21cm-1 absorption peak, in line of site to stellar objects at known distances. At an average H density of approximately 1/cm^3 ($\sim 1.67 \times 10^{-21}$ kg m^{-3}) within the interstellar vacuum of the Milky Way, it is very nearly collision-less and given its ground state, does not emit light. Whenever H exists above the ground state, and is significantly more concentrated, it is readily visible and termed a cold, or warm, or a hot gas cloud. Following a brief review of the historical evidence for the existence of dark matter, the key observations of the 21 cm^{-1} absorption peak reported during 2018-2019 are summarized, and the current constraints imposed on dark matter elaborated. The author's calculations and reasoning about why very cold H is probably "dark matter", in the context of these observations, are presented.

Michael L. Smith
Independent Researcher,
Umeå, Sweden

Chapter 1

Rethinking "Dark Matter" within the Epistemologically Different World (EDW) Perspective

Gabriel Vacariu and Mihai Vacariu

The really hard problems are great because we know they'll require a crazy new idea.
(Mike Turner in Panek 2011, p. 195)

Abstract

In the first part of the article, we show how the notion of the "universe"/"world" should be replaced with the newly postulated concept of "epistemologically different worlds" (EDWs). Consequently, we try to demonstrate that notions like "dark matter" and "dark energy" do not have a proper ontological basis: due to the correspondences between two EDWs, the macro-epistemological world (EW) (the EW of macro-entities like planets and tables) and the mega-EW or the macro–macro-EW (the EW of certain entities and processes that do not exist for the ED entities that belong to the macro-EW). Thus, we have to rethink the notions like "dark matter" and "dark energy" within the EDW perspective. We make an analogy with quantum mechanics: the "entanglement" is a process that belongs to the wave-EW, but not to the micro-EW (where those two microparticles are placed). The same principle works for explaining dark matter and dark energy: it is about entities and processes that belong to the "mega-EW," but not to the macro-EW. The EDW perspective (2002, 2005, 2007, 2008) presupposes a new framework within which some general issues in physics should be addressed: (1) the dark matter, dark energy, and some other related issues from cosmology, (2) the main problems of quantum mechanics, (3) the relationship between Einstein's general relativity and quantum mechanics, and so on.

Keywords: universe, dark matter, epistemologically different worlds, correspondence, clusters of galaxies, cosmic filaments, "standard" Lambda-CDM model of cosmology

1. The epistemologically different worlds (EDWs): principles concerning the existence and the interactions of "epistemologically different entities"

In our previous works, working on the mind-body problem, we demonstrated the existence of epistemologically different worlds (EDWs).[1] Later, we applied this perspective to the main problems of quantum mechanics (entanglement, nonlocality, etc.) and then to the relationship between micro-entities and macro-entities. We constantly believed that the greatest problems of particular sciences are philosophical problems that require a new framework of thinking.[2]

In this chapter, we rethink one of the most important notions in cosmology today, the "dark matter," within the EDW perspective. Therefore, let us introduce, very shortly in this section, the EDW perspective.[3] We will illustrate the principles referring to the existence of nonliving objects and their interactions, in general. The questions to start with are as follows. Do the micro-entities and the electromagnetic waves really exist? Do the macro-entities (and gravity) really exist? Within the EDW perspective, the main idea has been that the "universe"/"world" cannot even exist; what really exist are epistemologically different worlds (EDWs).

We introduce the five principles concerning physical objects and their interactions:[4]

1. Epistemologically different interactions constitute epistemologically different objects, and epistemologically different objects determine epistemologically different interactions.

2. Any object exists only at "the surface," due to the interactions that constitute it.

3. Any object exists in a single EW and interacts only with the objects from the same EW.

4. Any EW (a set of objects and their interactions) appears from and disappears into nothing.

5. Therefore all EDWs share the same objective reality, even if one EW does not exist for any other EDW ([7], pp. 25-26).

Every object exists in only one epistemological world (EW). It means that the object exists and interacts only with entities from the same EW. The electromagnetic waves, the microparticles, and the planets existed long before man appeared on the earth. The interactions of an entity constitute the surface of that object. The macro-objects interact among them; the micro-entities interact among them; and the electromagnetic waves interact among them. Essentially, a macro-entity does not exist for a micro-entity; an electromagnetic wave does not exist for either the micro-object or macro-object. There are only correspondences between ED entities that belong to the EDWs: a macro-object corresponds to a micro-object which corresponds to an electromagnetic wave. Obviously, all macro-objects exist in the macro-EW, and all

---

[1] In 2016, we tried to prove that in cosmology scientists deal pseudo-notions, such as dark matter, dark energy, the existence of space and time (or spacetime), inflation, and so on and so forth. For instance, taking into account the theory of simultaneous "Big Bang", the notion of "inflation" can be rejected (Alan Guth, etc.).

[2] Paul Dirac believed that the greatest problems of quantum mechanics were philosophical problems.

[3] For more details, see our previous works.

[4] These principles appeared in Gabriel Vacariu's previous works [1–6].

micro-entities exist in the micro-EW. A macro-object or a micro-entity exists just because it interacts with entities from the same EW. An electron exists just because it interacts with the microparticles from the same EW. An electron does not exist/interact for a table/planet but for an amalgam (which corresponds with that table/planet). Until we discovered the EDWs, physicists believed that a macro-entity is "identical" with an amalgam of microparticles. However, a table/planet is not identical with an amalgam of microparticles because the macro-entities and the micro-entities have different properties. For instance, we cannot reduce gravity to micro-entities. Also, we cannot reduce a microparticle to an electromagnetic wave.

In this context, we introduce our main assumption: *two objects cannot exist in the same place, at the same time*. Consequently, we cannot assume that a table and the corresponding amalgam of microparticles are "different (set of) entities in the same world," and there is no point in claiming that the microparticles "form" or "compose" a table or a planet. Composition, emergence, supervenience, and identity are all obsolete notions created within an obsolete framework—the "universe"/world, or what we called the "unicorn world." In this context, we can indicate that a planet appeared out of "nothing" (within the macro-EW) but this macro-EW corresponded to the micro-EW.

One of the greatest problems in the history of human thinking was the relationships between various "entities." "Causality" is one of the main problems in the history of human thinking. Causality is strong related to the "physical laws." Related to "causality" is the notion of "levels." It is meaningless to check for the causality between entities that belong to EDWs, since one EW does not exist for any EDW. We can talk about "causality" only between two entities that belong to the same EW, but not about causality that refers to entities that belong to EDWs! Searching for the "causalities" between the entities that belong to the EDWs has created many Ptolemaic epicycles during the entire history of human knowledge. Also, we have to mention here that some EDWs (the micro-EW or the macro-EW, for instance) do not really exist, that is, they do not have their ontologies; what really exist there are certain ED entities and their interactions that only *represent* those EDWs for us.

As observers, in order to observe (*indirectly*, through correspondences) the entities in a particular EW, we need certain conditions of observation. For example, we can observe an electron through a microscope (which can be regarded as a macro-object), but this is an indirect observation as the electron belongs to a micro-EW. The electron does not interact with the brain or the body of the researcher and not even with the microscope itself (a macro-object), but it interacts with an amalgam of microparticles that *corresponds* to that electronic microscope.

Through the processes, we observe entities belonging to EDWs are indirect and occur through *correspondence*, even in the case of macro-objects, not just for the micro-objects. We can change our conditions of observation in order to change observing indirectly EDWs. With our eyes, we can observe, indirectly (our mind-EW is involved) a table. By changing our conditions of observation (adding an electronic microscope), we can see, indirectly, an amalgam of microparticles which corresponds to that table. The table interacts with other macro-entities (a cup, a book placed on top), and this is the reason we consider that the table really exists. At the same time, an amalgam of microparticles that *corresponds* to a cup interacts with an amalgam of microparticles that corresponds to that table. In the world of microparticles, any macro-entity does not exist. In the world of electromagnetic waves, any microparticle or macroparticle does not exist!

In conclusion, the universe/world does not really exist but the EDWs do[5]. More exactly, the ED entities (like the macro-entities, the micro-entities, the

---

[5] We emphasis that EDWs are totally different than "parallel worlds/universes" from actual physics.

electromagnetic waves) and their interactions really exist in the EDWs. We repeat the main principle of EDW perspective: one EW does not exist for any EDW!

2. Rethinking "dark matter" within the EDW perspective

The most difficult problem of cosmology in our days is the dark matter and dark energy. What is dark matter? Contemporary thinkers believe that the gravity of:

> *[...] dark matter must therefore be the "glue" that holds galaxies like our own together. The fate of the universe itself seems to hinge on the total amount of dark matter and the properties of a mysterious form of energy—often called dark energy —that appears to be counteracting the effects of gravity on large scales ([8], p. 445). Dark matter is the name given to mass that emits no detectable radiation; we infer its existence from its gravitational effects... dark matter is the name we give to whatever unseen influence provides the gravity needed to explain the motions we observe. Dark energy is the name given to the unseen influence that may be causing the expansion of the universe to accelerate with time. ([8], pp. 446-447).*

There are direct and indirect methods of detecting the dark matter [9].[6] For instance, the amount of dark matter in a galaxy is determined by comparing the mass of the galaxy with its luminosity (mass-to-light ratio)[7]. The main problem is that plotting "the orbital speeds observed at different distances for most spiral galaxies shows that these speeds do not drop off with distance from the center (...)" [8]. It is believed that in the first billion years of the "universe," dark matter had no role even if it was present in that period. In fact, we consider that the EW of dark matter (the mega-EW) appeared when, in the macro-EW, the galaxies and the cluster of galaxies were formed. It means that if any galaxy was not formed in a particular place, then there was no EDW with the mega-entities that correspond to the galaxies. Probably, there are mega-entities that correspond to the individual galaxies, but there are also mega-entities that correspond to the "clusters of galaxies."

Let us introduce the chronological order of some people who have worked on the dark matter. Krauss [13][8] mentioned the names of some important people working

---

[6] Using the gravitational lensing method, Gilman et al. [10] detected the "existence" of cold dark matter. Their results are in concordance with the "predictions of cold dark matter." We emphasize that the results refer to the "sub-galactic scales." These sub-galactic scales refer, in fact, to the mega-entities that exist in the mega-EW. "At present, there's no direct evidence in the lab that dark matter particles exist," Birrer said. "Particle physicists would not even talk about dark matter if the cosmologists did not say it's there, based on observations of its effects. When we cosmologists talk about dark matter, we are asking 'how does it govern the appearance of the universe, and on what scales?'" [11]. Obviously, there are no "dark particles," and the scale is the mega-scale, that is, the mega-EW.

[7] "We can determine the amount of dark matter in a galaxy by comparing the galaxy's mass to its luminosity. More formally, astronomers calculate the galaxy's *mass-to-light ratio* (see Cosmic Calculations 16.1). First, we use the galaxy's luminosity to estimate the amount of mass that the galaxy contains in the form of stars. Next, we determine the galaxy's total mass by applying the law of gravity to observations of the orbital velocities of stars and gas clouds. If this total mass is larger than the mass that we can attribute to stars, then we infer that the excess mass must be dark matter" [8]. "There was clearly a discrepancy between the luminous mass observed with telescopes and the mass inferred from dynamical measurements (...)" ([12], p. 25).

[8] As many other physicists, Krauss [13] tries to show that the "Universe" appeared from "nothing" (even space and time). (For a short introducing to "Nothing" see also Close 2009) As we showed with the EDWs perspective, the universe/world does not exist but the EDWs are.

in cosmology in the first decades of the twentieth century[9] (but we added other persons on his list): Lord Kelvin who introduced the "dark bodies" and Poincare (1906) who used the term "dark matter" [15]; Lemaitre who proposed the Big Bang in the 1920s; Hubble, one of the most important astronomers:[10] together with Milton Humason, he proposed the "Hubble law" and radio astronomy pioneer [16].

Krauss mentioned a problem: "comparing with the abundance of light elements, the density of protons and neutrons produced by Big Bang should be doubled that it exists and consequently, it was necessary the introduction of "dark matter", something mysterious that flowed betweed the stars and ran the whole gravitational show we call a galaxy" ([13], p. 46).[11]

It has been supposed that the particles that produce the dark matter are weakly interacting massive particles (WIMPs), axion, neutrino, neutralino, or many other particles[12]. It is completely meaningless to search for the microparticles that compose the "dark matter" since the mega-entities belong to the mega-EW. Therefore, the microparticles do not exist for the dark matter, and the dark matter does not exist for any kind of microparticles! Obviously, from our EDW perspective, there are no "atoms" (microparticles) that "form" the dark matter. The movements of the galaxies (their masses) have to be regarded in relationship with other galaxies and not with the masses of planets (the macro-EW) to "form" the galaxies. Within the macro-EW, the galaxies do not have any ontological status but only the planets that represent, for us, the galaxies. The same principle is available for investigating the relationship between the microparticles and the macro-entities: *there are ED laws for ED entities*.[13] The microparticles *correspond* to the planets, so it would be wrong to consider that the microparticles "form" the planets. In consequence, it would be wrong to consider that the "planets form the galaxies"! Indeed, the "missing mass" is not "something else" in the macro-EW, but there are the mega-entities (the mega-EW) that correspond to the "galaxies" (the macro-EW)![14] In the mega-EW, the planets or the microparticles do not even exist; this EW has more ED entities and ED laws than the macro-EW or the micro-EW! From our viewpoint, dark matter does not exist within "our universe" at all! In fact, there are no "mysterious particles" that we cannot observe empirically since they do not exist in any

---

[9] For the history of a longer period of cosmology, see [14].

[10] "Hubble was able to use his measurement of Cepheids and Leavitt's period-luminosity relation to prove definitively that the Cepheids in Andromeda and several other nebulae were much too distant to be inside the Milky Way" ([13], p. 31).

[11] "Take it [dark matter] away from a galaxy like our own Milky Way, and all its stars and planets would fly away like bullets in intergalactic space!" ([12], p. vii). One of the great reasons that support the existence of dark matter is that the "orbital speeds" in the Milky Way are very high even if the stars are very far from the center of the galaxy [8]. In contrast, because the gravitational field of the sun decreases with its distance from our solar system, the orbital speeds decrease with this distance!

[12] "Physicists have proposed literally tens of possible dark matter candidates, including neutralinos, gravitinos, sneutrinos, sterile neutrinos, axions, fuzzy dark matter, WIMPs, WIMPzillas, superWIMPs, self-interacting dark matter, cryptons, Kaluza–Klein dark matter, D-matter, branons, Q-balls, and mirror matter, to name a few" ([12], p. 61).

[13] We emphasize that the expression "the ED entities" does not involve that one entity (from a particular EW) is the *sum* of other entities (from an EDW). For instance, a mega-entity that corresponds to a galaxy or cluster of galaxies has properties other than the properties of the planets (and their "empty space") that represent, for us, the "galaxy."

[14] Until we have written this paper, we used the notion "the macro–macro-EW," but based on Prof. Ilie Parvu's suggestion, we replaced this notion with "the mega-EW."

macro-EW[15] that we have discovered. There are only some phenomena that involve the ED entities and the ED interactions that belong to an EDW! These phenomena only *correspond* to some strange phenomena that cannot be explained in the macro-EW. The macro-EW that contains macro-objects, planets, and stars that "form," for us, the galaxies is not the largest macro-EW. In fact, "largest" has no meaning regarding the comparisons of EDWs. It is a notion that presupposes the notion of space (but in our book 2016, we indicated that space and time (or spacetime) cannot have any ontology. Moreover, an EW does not exist for any EDW, so the notion "the largest EW" has no meaning. A "galaxy" (no ontological status) in the macro-EW corresponds to an EDW, the mega-EW.[16] Essentially, a star appeared with other planets that "formed in a flattened disk surrounding it" ([8], p. ...). This idea is quite important in explaining the "dark matter" of a "galaxy."

Let us imagine a disk (a CD) threw in air by a human hand in an "empty space" (long distance from any planet and their gravitation). The CD will rotate exactly as a galaxy rotates. The margin of that disk rotates with a speed much greater than the speed of points closer to the center of the disk. The force acting on the disk (centrifugal force) corresponds to the micro-forces that bring together these micro-particles and their rotation even if we cannot understand the origin of this centrifugal force. In the micro-EW, there are the microparticles, their micro-forces, and their "dark" rotation. In the macro-EW, there is a disk with a centrifugal force (and maybe gravitational force). According to the principles of EDW perspective, the microparticles exist just because of their interactions within the micro-EW, and the macro-entities (stones, planets) exist because of their interactions within the macro-EW. The "dark matter" (the mega-entities) from the mega-EW corresponds to the planets and the empty space among them (which only represent, for us, the galaxies). Essentially, from the EDW perspective, *the mega-entities that represent the mega-EW exist just because of the mega-interactions between them within this EW.* A mega-entity does not exist for a planet (or a galaxy) (a planet does not exist for a mega-entity) just because their interaction is meaningless, since the mega-entity and the planet are ED entities that belong to EDWs which do not exist for the other![17]

---

[15] Not surprisingly (from our viewpoint), recent experiments for searching the dark matter particles furnished negative results: Large Underground Xenon (LUX) (Dakota); XENON1T, XENON10, and XENON100 (Italy); PandaX-II (China); and LHC (Geneva) found no evidence for dark matter particles ([17], p. 40)! Some researchers introduced the notion of "unseen particles"/forces ("hidden sector") [17]. Within the EDW perspective, the so-called unseen particles is a totally wrong notion! Quite interestingly, in 2016, McGauch et al., measuring the "gravitational pull" from "normal matter" of 150 galaxies with gravitational pull from dark matter, discovered a strong "correlation between dark matter and normal matter ([17], p. 41)! In fact, it is about the correlations between the ED entities that belong to two EDWs: the macro-EW (galaxies) and the mega-EW (the mega-entities). The correlation refers to two kinds of "gravity": the gravity of "galaxies" and the gravity of mega-entities.

[16] Moreover, there are other EDWs: for instance, Krauss informs us that "the largest gravitationally bound objects in the universe are called superclusters of galaxies. Such objects can contain thousands of individual galaxies or more and can stretch across tens of millions of light-years. Most galaxies exist in such superclusters, and indeed our own galaxy is located within the Virgo supercluster of galaxies, whose center is almost 60 million light-years away from us" ([13], p. 48). These superclusters of galaxies (which do not have any ontological status in the macro-EW) correspond to ED entities that belong to an EDW, the mega-EW.

[17] In general, there are two alternatives for the existence of dark matter: either "dark matter microparticles" or "changing the gravity equations" (initiated by Mordehai Milgrom in 1983 who modified Newton's laws, creating "modified Newtonian dynamics" (MOND)). Recent observations of gravitation in galaxies favor the modified gravity theories over dark matter ([17], p. 38).

In 1937, Zwicky proposed that using Einstein's method of gravitational lens, it is possible to test general relativity, to magnify more distant objects, and to find out why clusters appear to weigh more than what can be accounted for by visible matter ([13], p. 51).

More important is the observation about the dark matter "haloes"[18], "big blobs of dark matter in which galaxies were embedded—were necessary to keep the structures of many spiral galaxies stable"[19] ([14], p. 334). The "haloes" (no ontological status within the macro-EW) and the galaxies (no ontological status) formed by planets (macro-ontological status) correspond to the mega-entities that exist in the mega-EW.

How was each galaxy formed? The main force was gravitation that "acts and isolates clumps of matter on all scales" ([14], p. 334). This idea mirrors one of Gabriel Vacariu's main principles from his works ([2, 21], 2008, etc.): in this case, the main principle is "The interactions constitute the entities, the entities determine their interactions." According to the gravitation, we cannot explain the movements of planets that are at the margin of the disk: these planets have too much speed in relationship to gravitation. Our bodies (our eyes) are particular entities within the macro-EW where we can find the planets and their movements. We cannot see any "supersystem galaxies" since this "supersystem" is an entity (or maybe an amalgam of entities) that exists in the mega-EW. That mega-EW does not exist for our bodies, for planets and galaxies that we can observe, or for our minds since all these entities belong to EDWs.

One of the most important actual cosmologists regarding "dark matter" is James Peebles who mentions that it "might be the DM that gravitationally binds clusters of

---

The astronomers consider that all that exist are the galaxies and reject the existence of dark matter. They have worked within the unicorn world: the mega-entities really exist but in the mega-EW. In reality, Einstein's general theory of relativity is available only for planets (the macro-EW), but not for the mega-entities (the mega-EW)! So, we do not need to modify this theory; we have to discover the new laws governing these mega-entities!

[18] "By knowing the number of galaxies, cosmologists then estimate the amount of dark matter in the universe" ([18], p. 120). "Dark matter and light elements like hydrogen and helium were produced in the first few minutes after the Big Bang. Dark matter halos then slowly grew from seed structures and merged into ever-larger systems, until gas fell under their gravitational pull and sunk to their centers" ([12], p. 28). "The component of the galaxy that is *not* seen, because it is too diffuse, is the galactic halo, a spherical region of diameter so large that it encompasses the whole of the visible part of the galaxy. The stars within the halo are solely older Population II stars and many of these are within globular clusters. The most important component of the halo is what we cannot see — dark matter" ([19], pp. 115–116). "The studies found that dark matter surrounds most galaxies in roughly spherical clouds, called halos. Dark matter halos are significantly larger than the visible part of most galaxies, and often extend well into intergalactic space" ([20], p. 19). "Encompassing the Milky Way galaxy is a halo of dark matter. The particles making up this enormous dark matter cloud travel through every corner of our galaxy, oblivious of the planets, stars, dust, and other forms of ordinary matter around them. To a particle of dark matter, the world is a lonely and quiet place" ([20], p. 106). Indeed, to a "particle"/entity of dark matter, from what we know that exists, nothing exists! Therefore, there is only the mega-EW, and no other EDW exists for this world. We strongly emphasize that the "halo" is similar to "ether" in the end of the nineteenth century!

[19] "Looking towards the constellation Sagittarius, you'll be looking at the Galactic center, which is at the same time the center of the disk of stars and gas of our galaxy, which constitutes essentially everything you can see in the sky with the naked eye, and the center of a spheroid of dark matter, the halo, about ten times larger, and ten times more massive than the disk" ([12], p. 5). The "halo" is similar to the mental causation in philosophy of mind or graviton in physics. In fact, the haloes are nothing in the macro-EW but *correspond* to certain entities/processes that belong to the mega-EW, for instance.

galaxies[15,16]" ([22], p. 1)[20], but we have to be aware that the dark matter "does not bind clusters of galaxies" and the mega-entities that belong to the mega-EW correspond to the clusters of galaxies (planets and empty spaces among them). In 2015, Peebles writes about the "galaxy phenomenology," proposing the concept of "pure disk galaxies" in which "most of the stars move in streams in directions close to the plane of the disk, as in whirlpools and bars" ([24], p. 12248).

However, from the EDW perspective, the "disk galaxies" have no ontology [the galaxies are formed by planets, but these planets and the empty spaces among them correspond to the mega-entities (the mega-EW)]! Peebles's "galaxy phenomenology" sends directly to our hyper ontology of EDWs: it is about the mega-entity (a mega-disk) within the mega-EW.

More interestingly, in a paper from 2014, the entire Part 4 has the title "Island universes." Peebles concluded that "two broad classes of galaxies, pure disks and elliptical, have evolved in near isolation from their surroundings, as island universes" ([25], p. 10). From our viewpoint, Peebles needs the EDW perspective to provide the *ontology* of dark matter, namely, the ontology of "island universes": these are the mega-entities that belong to the mega-EW.[21] Cosmologists believe that:

> *Dark matter provides, in a way, the 'stage' for the 'cosmic show', a stage that was assembled when the universe was young, way before the time when stars started to shine and planets started to form, and this stage is still evolving. It is, in short, the supporting structure of the universe. It solves in a single stroke many problems in astrophysics and cosmology, and it provides a self-consistent framework for the structure and evolution of the universe. ([12], p. 4).*

We can make an analogy between a table and the corresponding amalgam of microparticles. The format of that amalgam of microparticles has no meaning: why this format has that shape? Within the micro-EW, we cannot find any meaning for the format of that amalgam of microparticles. However, everything gets a meaning if we introduce the *correspondence* between that amalgam of microparticles and the macro-table that belong to EDWs. Also, the galaxies have a particular format: their constituents (the planets) move with a particular speed just because they correspond to a "disk," a mega-disk.[22] If we rotate a disk in the macro-EW, a second person, using a microscopic electron, will observe an amalgam of microparticles that is arranged under a "disk format," where all microparticles moving with the same speed! So, we can presuppose that, because of the Big Bang and other

---

20 "... Jim Peebles had pointed out that the absence of fuctuations in the cosmic microwave background at a level of ~10^-4 was incompatible with a Universe that was composed of only baryonic matter, and argued that this problem would be relieved if the Universe was instead dominated by massive, weakly interacting particles, whose density uctuations could begin to grow prior to decoupling (239) (see also, Ref. [79])" ([23], p. 58).

21 For instance, Peebles writes that "How could the progenitor fragments of pure disk galaxies have 'known' not to have participated in this generally high global star formation rate? One piece of the matter tumbling together according to the ΛCDM picture of the formation of the pure disk galaxy in Figure 3 'knew' it was going to host the growing disk, and start growing it at redshift well above unity if the age of the disk of the Milky Way [11] was typical of pure disk galaxies, while the rest of the fragments 'knew' they had to hold off star formation until they had reached the growing disk. It is a curious situation" ([25], p. 8). It is a "curious situation" within the unicorn world (the universe); however, within the EDW perspective, that problematic notion, "knew", has a meaning: the growing disk corresponds to a mega-entity (the mega-EW).

22 "Spiral arms are waves of star formation that spread through our galaxy's disk" [8].

phenomena, billions of planets of a galaxy have been moving under the format of a disk, all planets having the same speed. These "galaxy disks" correspond to the mega-entities, the mega-disks (the mega-EW)!

Working within the unicorn world, the physicists logically believe that dark matter does not "emit or absorb electromagnetic radiation" (it is "dark") and does not have any kind of interactions with the "known matter" ([14], p. 334). Again, dark matter cannot interact with anything from the macro-EW (in which there are planets that form, for us, the galaxies, for instance); it cannot emit or absorb electromagnetic radiation, since it does not exist for the ED entities and ED forces that belong to EDWs. Anyway, working within the unicorn world, many scientists believe that dark matter does not interact with any kind of matter that we know,[23] but it is impossible for us to see the causes of such strange phenomena. Hooper claims that the dark matter is not just "out there" but it is everywhere, in our world, and at the same time, this "new type of elementary particles" does not exist ([20], p. v). Also, there is no "direct influence" or any kind of "interactions" between the dark matter (the mega-entities that belong to the mega-EW) and any kind of matter that belongs to EDWs.

There are the macro-EW, the micro-EW, the wave-EW, the mind is an EW, therefore, there has to be the mega-EW, an EW, in which there is the "matter" (the mega-matter) that corresponds to the *indirect effects* (i.e., through *correspondences*) in the macro-EW and the macro-entities like planets that form, for us, the galaxies and the clusters of galaxies. The dark matter has to be a kind of nonbaryonic matter since any star is formed from baryonic, normal matter.[24] There are no interactions between baryonic matter and nonbaryonic matter since one kind of matter does not exist for the other kind of matter. The amount of dark matter in a galaxy is determined by comparing the mass of galaxy with its luminosity (mass-to-light ratio). "The evidence of dark matter is by and large gravitational. The discrepancy between the luminous mass and the gravitational mass gives an indication of the presence of a huge unseen mass in the Universe" ([27], p. 89). Dark matter has an *indirect* influence on the "empty space," but this "nothing" corresponds to "something" that belongs to the mega-EW!

Exactly as an electron does not interact with a planet but with an amalgam of microparticles, the dark matter does not exist for the macro-objects (like planets). The galaxies (the planets and the space among them) correspond to an entity that belongs to the mega-EW. Nothing can stop us to introduce this idea. The human body is placed between the microparticles and the galaxies, but we can push further the dimension of certain entities: these are the mega-entities that have "greater" dimensions than the macro-objects. Just as macro-observers, we cannot perceive/understand the rotation of a "galaxy" from the viewpoint of a mega-entity (mega-entity) since the mega-entity

---

[23] Similar ideas have been invented for the explanations of the "entanglement" in quantum mechanics or the "mental causation" in philosophy of mind. Using the EDW perspective, in our previous works, we indicated that entanglement, nonlocality, and many other notions from quantum mechanics are pseudo-notions constructed within the unicorn world, the universe! (For these pseudo-notions, see our previous works). "As I have mentioned above, dark matter particles are all around us-in the room in which I am typing, as well as 'out there' in space. Hence we can perform experiments to look for dark matter and for the new type of elementary particle or particles of which it is comprised" ([13], p. 54). For us, Krauss' idea mirrors the "correspondences" between entities, phenomena, and forces that belong to EDWs.

[24] The "dark matter cannot consist of normal matter made up of neutrons and protons; if it did, the density of neutrons and protons in the early universe would have been much higher, and the resulting abundances of light elements in the universe would have been much different from what we actually observe" ([26], p. 376). Again, the dark matter (the mega-matter) corresponds to the planets that correspond to the microparticles (neutrons and protons) that correspond to the electromagnetic fields.

does not exist for the planets that form the galaxies! Most probably, the rotation of a "galaxy" *corresponds* to the rotation of a mega-entity (the mega-EW).

Today, there are several reasons for supporting the Big Bang, the phenomenon that did take place approximately 13.78 or 13.82 billion years ago. From our viewpoint, exactly as the gravity does not exist for the electron (there are no "gravitons"), the indirect effects of gravitation exist for the microparticles.[25] What is important is that cosmologists believe that a star appeared with other planets that "formed in a flattened disk surrounding it" ([8], p. ...). This idea mirrors exactly the existence of the mega-entities. A galaxy (no ontology) (formed by planets with ontological status in the macro-EW and the empty space among them) corresponds to a mega-entity that belongs to the mega-EW. Exactly as an electron cannot "perceive"/interact with a table (because the table does not exist for the electron), we cannot perceive/interact with a mega-entity. The mega-entity rotates exactly as a macro-disk rotates in the macro-EW. With external limits, the disk rotates with much greater speed than its center. This analogy is very approximate because the spiral galaxies are not spinning similar to the solid bodies and they do not mimic the motion of the planets around the sun, where velocity decreases with distance ([28], p. 21). The "disk" in the mega-EW is not exactly like a disk in the macro-EW: there are different properties of these two disks (the macro-disk and the mega-disk), but we are unable to identify, directly, the properties of the mega-disk. We will be able to identify these properties only indirectly since our bodies are macro-entities that do not exist for the mega-entities. In 2007, writing about Kant's philosophy, Gabriel Vacariu concluded that within the EDW perspective, the galaxies are entities different from tables, stones, or even individual planets, and exactly as an electron "does not exist" in a macro-EW, a planet "does not exist" in a macro–macro-EW ([29], p. 17). There are no "causations" that would require direct relationships between the ED entities that belong to two EDWs since the entities from an EW do not exist for the entities that belong to an EDW. From indirect observations, we can conclude that the "dark matter" really exists but in the mega-EW.

We return to our analogy between a macro-disk and the corresponding amalgam of microparticles: if a micro-observer observes the rotation of an amalgam of micro-particles (without being able to observe the macro-disk), then that micro-observer would introduce certain "dark matter" for explaining the rotation of the microparticles. For the micro-observer, the macro-disk cannot even exist! We can continue the analogy introducing the rotation of a planet which corresponds to a huge amalgam of microparticles. The micro-observer would need to introduce dark matter/energy for explaining the rotation of that amalgam of microparticles! In this context, we make an important analogy regarding the relationship between "gravity and microparticles" and the relationship between "dark matter/energy and macroparticles":

> *Gravity (the curvature of spacetime that "belongs" to the macro-EW) for microparticles that belong to the micro-EW is quite similar to dark matter and dark energy that belong to the mega-EW for the macro-entities that belong to the macro-EW.*

---

[25] In his PhD thesis and his first book, Gabriel Vacariu [5] indicates that "gravity" does not exist as a force (Newton) or as a "curved spacetime" (Einstein) but as "nothing" in the macro-EW (no ontological status!) which *corresponds* to the curved electromagnetic fields that belong to the field-EW. In other words, it is the electromagnetic field that is *indirectly* "curved," and the "curvature" is not produced by the planet (which does not exist for any electromagnetic field) but by a huge amalgam of electromagnetic waves (field-EW) which corresponds to a huge amalgam of microparticles (the micro-EW) which corresponds to the planet (the macro-EW). In our book (2016), we indicated that "spacetime" cannot have any ontological status—it would produce strong ontological contradictions; in 2017, we rewrote Einstein's both special and general theories of relativity not using "spacetime" (which has no ontological status) but the *motions* of ED entities that belong to the EDWs!

A microparticle (a photon, for instance) does not "perceive"/interact with a planet; therefore, gravity does not exist for the photon. However, in its trajectory, the photon follows the "curvature of spacetime" produced by a planet/galaxy. The photon would "think" "It has to be a *dark matter, a dark halo* that surrounds this huge amalgam of microparticles!" The photon cannot even "perceive" that the spacetime is curved. We can think that there is a halo of dark matter that surrounds a galaxy, but exactly in the same way, an electron that moves around the proton would ask about certain "gravitational force," a planet would ask about the "dark matter" that surrounds a galaxy. Exactly as the "gravitation" does not exist for photons, dark matter does not exist for planets (and their galaxies). However, the photons follow the spatiotemporal paths (curved space) between planets, even if a planet does not exist for a photon. From the viewpoint of photons, we can think of certain microparticles ("gravitons") that produce this curvature, but the gravitons do not really exist. In the same way, the galaxies are "biased" with respect to the dark matter,[26] but the dark matter does not exist for the planets.[27]

Within the EDW perspective, what does it mean by the "density" of dark matter? It seems that there are some entities/interactions that belong to an EDW, an EW does not exist for any EDW, and therefore, the density of dark energy is constant. Between entities and processes that belong to the EDWs are just correspondences and these

[26] In order to explain the existence of dark matter, some researchers try "to explain the nature of the galaxy", and for this, "they are trying to redefine gravity ... We need dark matter in order to grasp how galaxies work". "Martin Kunz, an astrophysicist at the University of Geneva, explains that the structures of the Universe just could not function on a huge cosmological scale without dark matter. The current best cosmological model also depends on it: the so-called Lambda-CDM model, also known as the Standard Model. Using just a few parameters, it describes the development of the Universe since the Big Bang. It can explain important observations, such as the Universe's accelerating expansion rate, the cosmic microwave background, or the honeycomb-like distribution of galaxies with enhanced clusters of galaxies linked by thin, thread-like structures with vast empty spaces between them – the so-called voids" (idem). In fact, the galaxies and the so-called voids do not have any ontological background, but they correspond to the mega-entity that belongs to the mega-EW! Oliver Müller (Strasbourg) found dwarf galaxies in the constellation of Centaurus moving on a plane, all in the same direction around the central galaxy Centaurus A. They were not distributed randomly either, as is predicted by the large cosmological simulations using the standard model. Müller's subsequent article, published last year in the specialist journal 'Science', caused quite a stir. The distribution of galaxies such as those of Centaurus is still allowed in the Lambda-CDM model, but it predicts that only one out of a thousand galaxies could have such a structure. The problem is that the same phenomenon can be seen in our own local group of galaxies, both in the Milky Way and in the Andromeda galaxies. "If the three closest galaxies have to be regarded as outliers, then something cannot be right about the basic assumptions found in the Standard Model," says Müller. However, he does not simply assume that the whole standard model is wrong, because it offers too many observations of the universe that are correct. Müller is simply pointing out certain discrepancies between his observations and the simulations of dark matter. "Perhaps we are missing something additional in the simulations," he says. "It's also possible that our galactic neighbors are just very special'. This is because the Milky Way, the Andromeda Galaxy and Centaurus A all lie on the edge of a huge void, and have the Virgo cluster of galaxies in direct proximity. Their mass distribution could thus lead to unusual phenomena". The standard model maybe is available for the macro-entities, but not for the mega-entities! In reality, the galaxies and the cluster of galaxies correspond to certain mega-entities that belong to the mega-EW.

[27] "It is possible that dark matter may have its own rich phenomenology hidden from the ordinary matter. This hidden dark matter sector might possess new forces and particles, some of which could be viable dark matter particles that are strongly self-interacting2" ([30], p. 3). In reality, dark matter belongs to the mega-EW which does not exist for the macro-EW; therefore, it is not about a "hidden" dark matter but an EDW!

correspondences are always constant since, for instance, in the macro-EW, where there are the "galaxies" (no ontological status), planets, and "nothing" and all these correspond to "something" that belong to an EDW (the mega-EW, for instance)! It has to be clear that the dark matter/energy belongs to an EDW rather than to the micro-EW (microparticles), the macro-EW (planets), and the field-EW (electromagnetic fields/waves); therefore, it is meaningless to check for the interactions between the dark matter and planets, microparticles, and electromagnetic waves.[28]

3. Recent cosmological results which strongly support the being of the mega-EW

In a recent article, Hutsemékers et al. indicated that the "quasar spin axes are likely parallel to their host large-scale structures" ([32], p. 1).[29]

Assuming that quasar polarization is either parallel or perpendicular to the accretion disk axis as a function of inclination, as observed in lower luminosity AGN, and considering that broader emission lines originate from quasars seen at higher inclinations, we inferred that quasar spin axes are likely parallel to their host large-scale structures. Galaxy spin axes are known to align with large-scale structures such as cosmic filaments. Till now, such alignments are detected up to redshift $z \sim 0.6$ at scales ≤ 100 Mpc.[30]

---

[28] A team from MIT realized certain experiments to detect the axions that would form the dark matter: "The team reports that in the first month of observations, the experiment detected no sign of axions within the mass range of 0.31 to 8.3 nanoelectronvolts. This means that axions within this mass range either do not exist or they have an even smaller effect on electricity and magnetism than previously thought" [31]. Within the EDW perspective, such experiments are meaningless: the dark matter does not exist for electricity or magnetism at all. "While they are thought to be everywhere, axions are predicted to be virtually ghost-like, having only tiny interactions with anything else in the universe" [31]. Quite wrong, dark matter is not composed of microparticles like axions, and moreover, there are not "only tiny interactions with anything else in the universe." On the contrary, the dark matter (the mega-entities) exists only in the mega-EW, but it does not exist for anything else in the EDWs! However, in the same article, there is an essential paragraph: "'As dark matter, they should not affect your everyday life,' Winslow says. 'But they are thought to affect things on a cosmological level, like the expansion of the universe and the formation of galaxies we see in the night sky'" [31]. Winslow (the principal investigator of the experiment) is quite correct, but she is missing the EDW perspective! In 2018, the researchers from MIT, using a magnetar, tried to detect the axions (the ABRACADABRA experiment). "The team proposed a design for a small, donut-shaped magnet kept in a refrigerator at temperatures just above absolute zero. Without axions, there should be no magnetic field in the center of the donut, or, as Winslow puts it, 'where the munchkin should be.' However, if axions exist, a detector should 'see' a magnetic field in the middle of the donut" [31]. Obviously, the results were negative: there are no microparticles that compose the dark matter (the mega-entities) since the microparticles and the mega-entities belong to the EDWs!

[29] In 2008, the astronomers from the University of Colorado Boulder indicated that they found the missing normal matter (baryons) in the spaces between galaxies. "'We think we are seeing the strands of a web-like structure that forms the backbone of the universe," said CU-Boulder Professor Mike Shull. 'What we are confirming in detail is that intergalactic space, which intuitively might seem to be empty, is in fact the reservoir for most of the normal, baryonic matter in the universe'.. The team also found that about 20 percent of the baryons reside in the voids between the web-like filaments. Within these voids could be dwarf galaxies or wisps of matter that could turn into stars and galaxies in billions of years, said the CU-Boulder researchers" (University of Colorado Boulder).

[30] "Likewise, in 1989 Margaret Geller and John Huchra, analyzing redraft survey data, discovered the immense 'Great Wall,' a 'sheet' formed from galaxies many light years apart. That first large-scale

Since coherent orientations of quasar polarization vectors, and then quasar axes, are found on scales larger than 500 Mpc, our results might also provide an explanation to the very large-scale polarization alignments reported in Papers I–III. In this case those alignments would be intrinsic, not due to a modification of the polarization along the line of sight. The existence of correlations in quasar axes over such extreme scales would constitute a serious anomaly for the cosmological principle (Hutsemékers et al., p. 5)[31].

Maybe, the "host large-scale structure" or "cosmic filaments" mirror the existence of the mega-EW. However, if these "cosmic filaments" refer to "intergalactic gas filaments" (baryonic matter), then it is not about the mega-EW. In principle, the mega-entities (the mega-EW) cannot be directly observed by the humans and their macro-tools! Anyway, the "cosmological principles" have to be changed, since the "universe/world" does not exist but the EDWs do. The scientists have noticed that some "galaxies" move together in odd and often unexplained patterns, as if they are connected by a vast unseen force. It is supposed that the dark matter was less influential in the first period after the "Big Bang." Ferreira considers that there is a sort of an influence of the so-called large-scale structures which influence the interactions between distant galaxies, structures made of hydrogen gas and dark matter, and take the form of filaments, sheets, and knots that link galaxies in a vast network called the cosmic web [34, 35].

Nevertheless, this "unseen force" has to be some entities or processes that belong to the mega-EW, but we are unable to notice them because they do not exist for the macro-entities (for our bodies and our instruments of observation, for instance). The "cosmic web" has to be something that belongs to the mega-EW, but not to the macro-EW. We emphasize again that the galaxies have no ontological status in the macro-EW but they correspond to the mega-entities that belong to the mega-EW.

Again, all these statements support the existence of certain mega-entities/processes that belong to the mega-EW. The secret of the "synchronized galaxies" is the existence of EDWs, i.e., the existence of mega-entities that belong to the mega-EW. Obviously, the wave-EW, the micro-EW, and the macro-EW really are. Because of the same reasons, the mega-EW should exist. *If the dark matter (mega-matter) really exists*,[32] then this matter exists in the mega-EW and have *indirect influence* (through *correspondences*) on the macro-entities and the macro-processes (the trajectories of planets, for instance) that belong to the macro-EW.[33]

---

structure is 500 million light-years long, 200 million light years wide, and with a thickness of 15 million light years" [33].

[31] The secret of these synchronized galaxies may indeed question the main cosmological principle that the universe is uniform and homogenous at extremely large scales, as Fereeira points out. He also mentions the work of Hutsemékers and his colleagues regarding "the correlations in quasar axes over such extreme scales." Furthermore, he considers that one of the most contentious debates in cosmology these days is centered around the unexpected way in which dwarf galaxies appear to become neatly aligned around larger host galaxies such as the Milky Way ([34], his highlights). It seems that not only a galaxy corresponds to a mega-entity but there are some mega-laws that involve these mega-entities. Obviously, since the "universe" does not exist, the old "cosmological principle" fails: the EDWs are not "uniform and homogenous" since one ED does not exist for any EDW! Here, it is about the structure of the mega-EW, not of the macro-EW (where large groups of planets form the "galaxies," for us, the observers).

[32] See Powell (2019) if dark matter really exist…

[33] Several groups of researchers which investigations have led to the conclusion that dark matter and dark energy do not exist at all! For instance, in "November, astronomers at the Chinese Academy of Sciences in Beijing published a paper identifying 19 galaxies which might violate the most fundamental theory of how the universe first formed. They had been searching the sky for yet-undiscovered galaxies which seem to be lacking the usual dark matter component, aiming to add more evidence to a baffling

The "standard" Lambda-CDM model of cosmology is quite accepted today: the total mass energy of the "universe" is 5% ordinary matter and energy, 27% dark matter, and 68% dark energy.[34] Obviously, this idea is constructed within the unicorn world! We strongly emphasize again that the "matter" from the micro-EW does not exist for the "matter" from the macro-EW and the matter from the macro-EW does not exist for the matter from the mega-EW! (the same idea is available for "energy" and "mater"!). Therefore it is meaningless to check for the microparticles that form the "dark matter"![35] An electron will never be able to interact with a planet just because the planet does not exist for the electron! The reader trying to discover dark matter has to imagine as being a photon (the micro-EW) searching the reason of its "curbed trajectory" near a huge amalgam of microparticles (which

---

phenomenon scientists had begun observing last year. And they claimed to have found a whole group of them" [36]. Van Dokkum and his team identified several galaxies without dark matter at all [36]! Also, Go Ogiya (Observatoire de la Côte d'Azur in France) "suggests a process by which galaxies might form without ever containing dark matter. So-called 'tidal dwarf galaxies' could form when dark matter and baryonic matter is 'ejected' from an existing galaxy due to tidal forces, but the dark matter component evaporates due to its higher velocity, leaving only stars and gas to form a new galaxy" [36]. "It may be that we do not yet fully understand how matter and energy evolved over time, particularly at early times," says his colleague Sherry Suyu, from the Max Planck Institute for Astrophysics, in Germany [37]. "Nicolas Martin, a researcher at the Observatory Astronomical De Strasbourg in France, believes that the observations needed to drive research forward are just beyond the limits of what is possible with the best apparatus around at the moment. Emailing while on location at one such cutting-edge telescope, he said that the research community would likely have to wait for two next-generation telescopes, currently planned or under construction in Chile and Hawaii, before they could generate even more precise measurements of the velocity of stars in the dwarf galaxies" [36]. Obviously, we will need more investigations regarding the existence of dark matter and dark energy, but we furnish this very recent information about dark matter and dark energy just to indicate that the framework necessary for explaining these processes is our EDW perspective: if dark matter really exists, it belongs to an EDW (the mega-EW) rather than to the EDWs that we already know. If the dark matter does not exist, it is very possible other mega-matter/processes will be discovered in the future, and this matter/process would belong to the mega-EW but not to the macro-EW! Indeed, being constructed under the wrong framework (the "universe"/world), cosmology of our days (physics, in general) has been in a crisis exactly as quantum mechanics has been in the last 100 years (with its great problems constructed within the wrong framework—the "universe", as we called the "unicorn-world"). We have solved all great problems of physics of the last century replacing the wrong framework, the "universe"/"world," with a much better framework, the EDW perspective!

[34] The standard ΛCDM cosmology assumes the general theory of relativity. This is an extrapolation of some 14 orders of magnitude in length scale from the precision tests on the scales of the Solar System and smaller. It assumes that 95% of the present mass of the universe is in two hypothetical forms, dark matter and dark energy" ([34], p. 1). "Observations over the past decades | obtained by combining a variety of astrophysical data, such as type-Ia supernovae, cosmic microwave background (CMB), baryon oscillations and weak lensing data | indicate that most of our Universe energy budget consists of unknown entities: 27% is dark matter and 68% is dark energy, 1 a form of ground-state energy" ([26], p. 1). (About these percentages, see also Panek 2011).

[35] "Although the existence of dark matter is generally accepted by the scientific community, some astrophysicists, [38] intrigued by certain observations which do not fit the dark matter theory, [20] argue for various modifications of the standard laws of general relativity, such as modified Newtonian dynamics, tensor-vector-scalar gravity, or entropic gravity. These models attempt to account for all observations without invoking supplemental non-baryonic matter. [17]" ([39], "Dark matter"). The idea of modifying the standard laws of general relativity is totally wrong! In reality, in order to explain dark matter, the physicists have to change their framework of thinking (the macro-EW) with the mega-EW!

corresponds with a planet in the macro-EW). Its curbed trajectory is due to the gravity of the planet, but the planet does not exist for the photon.

With the EDW perspective (2002, 2005, 2007, 2008), we generated the new framework of a new *Philosophiae Naturalis* necessary for (1) the main problems of quantum mechanics of the last 100 years and (2) the relationship between Einstein's general relativity and quantum mechanics, and (3) we furnished a new explanation of dark matter/energy (which presupposes the existence of mega-entities that belong to the mega-EW) and (4) many problems of cosmology today introducing the missing ED ontologies for many ED entities that belong to the EDWs! (see our previous works). The real great problems are hard not only because they require a "crazy new idea" (see the motto) but they require a new paradigm of thinking. Dark matter and many other problems of cosmology today (physics, in general) require the replacement of the "universe" (the "unicorn world") with our new paradigm of thinking, the EDW perspective![36]

---

[36] If you reject the existence of mega-entities (the mega-EW), then you also have to reject the existence of the macro-entities (the macro-EW). It means if your brain (a macro-entity) does not exist, then your mind (an EW which corresponds to your brain/body) cannot exist! So, your "rejection" (a statement that is a thought, a mental state, anyway) would be meaningless. A huge amalgam of neurons cannot produce thoughts but only chemical and electrical reactions (for the mind-brain problems, see [2–7]).

Author details

Gabriel Vacariu[1] and Mihai Vacariu[2]*

1 Department of Philosophy, University of Bucharest, Romania

2 National University of Political Studies and Public Administration, Bucharest, Romania

*Address all correspondence to: mvacariu@yahoo.com

References

[1] Vacariu G. The Epistemologically Different Worlds Perspective and Some Pseudo-Notions from Quantum Mechanics. Analele Universitatii Bucuresti; 2006

[2] Vacariu G. Mind, brain and epistemologically different worlds. In: Synthese Review. USA; 2005

[3] Vacariu G, Vacariu M. Physics and Epistemologically Different Worlds. Revue roumaine de philosophie. 2009, 2009;**53**:1-2

[4] Vacariu G. Kant, philosophy in the last 100 years and an epistemologically different worlds perspective. Revue Roumanie de Philosophie. 2007:51

[5] Terhesiu D, Vacariu G. Brain, mind and the perspective of the observer. Revue Roumanie de Philosophie. 2002; **46**:1-2

[6] Vacariu G, Terhesiu D. Brain, mind and the role of the observer. In: Botez A, Popescu B, editors. Philosophy of Consciousness and Cognitive Science. Bucharest: Cartea Romaneasca; 2002

[7] Vacariu G, Vacariu M. Dark Matter and Dark Energy, Space and Time, and Other Pseudo-Notions in Cosmology. Datagroup-Int, S.R.L.; 2016

[8] Jeffrey BO, Megan DO, Nicholas S. The Cosmic Perspective. 6th ed. Addison-Wesley; 2010

[9] Vasiliki MA. Dark matter: experimental and observational status. IFIC/19-17, Cornell University. 2019. Available at: https://arxiv.org/abs/1903.11589

[10] Daniel G, Birrer S, Nierenberg A, Treu T, Xiaolong D, Benson A. Warm dark matter chills out: Constraints on the halo mass function and the free-streaming length of dark matter with eight quadruple-image strong gravitational lenses. Monthly Notices of the Royal Astronomical Society. 2019; **491**(4):6077-6101. DOI: 10.1093/mnras/stz3480

[11] Hubble's Team. Hubble detects smallest known dark matter clumps. 2020. Available at: https://www.nasa.gov/feature/goddard/2020/hubble-detects-smallest-known-dark-matter-clumps

[12] Gianfranco B. Behind the Scenes of the Universe - from the Higgs to Dark Matter. Oxford University Press; 2013

[13] Krauss L. The Universe from Nothing – Why there Is Something than Nothing. Simon & Schuster; 2012

[14] William S, Christopher CJ. Galactic Encounters - Our Majestic and Evolving Star-System, from the Big Bang, to Time's End. Springer; 2015

[15] Stephanie BM. A history of dark matter. 2017. Available at: https://arstechnica.com/science/2017/02/a-history-of-dark-matter/

[16] Einasto J. Dark Matter. Braz J Phys. 2013;**43**:369-374

[17] Sabine H, Stancy MGS. Is dark matter real? Scientific American. 2018

[18] John MW. Reinventing Gravity - A Physicist Goes beyond Einstein. HarperCollins e-books; 2008

[19] Michael WM. Time, Space, Stars and Man - the Story of the Big Bang. 2nd ed2013

[20] Dan H. Dark Cosmos – In Search of our universe's Missing Mass and Energy. Harper Collins e-books; 2006

[21] Vacariu G. Epistemologically Different Worlds. Editura Universitatii din Bucuresti; 2008

[22] Peebles P, James E. Growth of the nonbaryonic dark matter theory. Nature Astronomy. 2017, 2017;**1**:art. 0057

[23] Bertone G, Hooper D. A history of dark matter. 2016. FERMILAB-PUB-16-157-A, arXiv:1605.04909v2 [astro-ph.CO] May 24, 2016

[24] Peebles P, James E. Dark matter. PNAS. 2015;**112**(40):12246-12248

[25] Peebles P, James E. The natural science of cosmology, proceedings of the 7th international conference on gravitation and cosmology (ICGC2011) IOP publishing. Journal of Physics: Conference Series. 2014;**484**:012001. DOI: 10.1088/1742-6596/484/1/012001

[26] Stacy P, Laura K, Brad S, George B. Understanding Our Universe. W. W. Norton & Company, Inc.; 2012

[27] Debasish M. Dark Matter - An Introduction. CRC Press Taylor & Francis Group; 2015

[28] Alain M, Vincent LB. Matter, Dark Matter, and Anti-Matter - In Search of the Hidden Universe. Springer; 2009

[29] Vacariu G. Epistemologically different worlds [PhD thesis]. Sydney, Australia: University of New South Wales, School of Philosophy; 2007. Available at: https://www.unsworks.unsw.edu.au/primo-explore/fulldisplay?vid=UNSWORKS&docid=unsworks_5143&context=L

[30] Zavala J, Carlos FS. Dark matter haloes and subhaloes. 2019. arXiv: 1907.11775v1 [astro-ph.CO] [Accessed: 26 July 2019]

[31] Chu J. Dark matter experiment finds no evidence of axions. 2019. https://www.sciencedaily.com/releases/2019/03/190328150940.htm

[32] Hutsemékers D, Braibant L, Pelgrims V, Sluse D. Alignment of quasar polarizations with large-scale structures. Astronomy & Astrophysics. 2018

[33] Robby B. Is the universe controlled by gigantic structures? 2019. Available at: https://bigthink.com/large-scale-structures

[34] Ferreira B. Cosmic web: Growing evidence that the universe is connected by giant structure. 2019. Available at: https://www.sott.net/article/423867-Cosmic-web-Growing-evidence-that-the-universe-is-connected-by-giant-structures

[35] Ferreira B. There's growing evidence that the universe is connected by giant structures. 2019. Available at: https://www.vice.com/en_us/article/zmj7pw/theres-growing-evidence-that-the-universe-is-connected-by-giant-structures

[36] Minsky C. Strange dark matter findings could rewrite the universe's history. 2020. Available at: https://www.vice.com/en_ca/article/pkedvy/strange-dark-matter-findings-could-rewrite-the-universes-history?fbclid=IwAR04yKJgFNG_D9FyVU22YGsE8N52KDH4uqcylb6OAm8czkodY7NiqM9ohoQ

[37] Richard LA. The Hubble Constant is constantly perplexing. 2020. Available at: https://cosmosmagazine.com/space/the-hubble-constant-is-constantly-perplexing?fbclid=IwAR08qDR1onyzRyroDxGHzs8xymfvVCE9zUEJtoUNZM4ska9lnhCtrATdH-k

[38] Paul H. Edge of the Universe – A Voyage to the Cosmic Horizon and beyond. John Wiley and Sons, Inc.; 2012

[39] Wikipedia. Dark matter. 2020. Available at: https://en.wikipedia.org/wiki/Dark_matter

Chapter 2

Bayesian Deep Learning for Dark Energy

Celia Escamilla-Rivera

Abstract

In this chapter, we discuss basic ideas on how to structure and study the Bayesian methods for standard models of dark energy and how to implement them in the architecture of deep learning processes.

Keywords: cosmology, dark energy, Bayesian analyses, machine learning, cosmological parameters

1. Introduction

The dark sector of the universe has been the issue of study for cosmologists who are striving to understand the world around us in its entirety. The composition of the current universe is an age-old inquiry that these researchers have probed into. And while we do have estimates of the likely percentages of baryonic matter, dark matter, and dark energy at 5, 27 and 68%, respectively, researchers have been trying to improve these estimates and optimise the computational expense of the statistical methods employed to analyse cosmological data available.

These thoughts have opened the path of the following chapter, in where we will discuss from the standard dark energy models to explain the cosmic acceleration until the design of a numerical architecture in order to understand the constrains over the cosmological parameters that can describe the current universe and its effects.

2. Dark energy as a solution to the cosmic acceleration

A highlight in observational cosmology is the origin and nature of the cosmic accelerated expansion. The standard cosmological model that is consistent with current cosmological observations is the so-called concordance model or Λ CDM. According to this scenario, the observed accelerating expansion is related to the repulsive gravitational force of a Cosmological Constant Λ with constant energy density ρ and negative pressure p. This proposal has been the backbone of the standard cosmology since the nineties, but simple enough as it is the proposal that has a couple of theoretical problems; two of them are the fine tuning argument and coincidence problem [1, 2]. In order to solve or at least relax these problems, some proposals have led to alternative scenarios that can modify the general relativity (GR) or consider a landscape with a dynamical dark energy. It is in this way that dark energy emerges as a cosmological solution since it can be described as a fluid

IntechOpen

parameterised by an equation of state (EoS), which can be written in terms of the redshift, $w(z)$. So far, the properties of this EoS remain under-researched. Just to mention a few, there are a zoo of proposals on dark energy parameterisations discussed in the literature (see, e.g., [3–9]), addressing from parameterisation as Taylor-like series to dynamical $w(z)$ that can provide oscillatory behaviours [10–13].

Nowadays, the techniques to discriminate between models and confront them with Λ CDM are based on the calculations of the constraints on the EoS-free parameter(s) of the models. This methodology has been done using observables that can show the cosmic acceleration such as supernovae type IA (SNeIa), baryon acoustic oscillations (BAO), cosmic microwave background (CMB), weak lensing spectrum, etc. The relevance of using these observations is due to the precision with which dark energy can be probed. Currently, some measurements such as the Pantheon from supernovae [14], BOSS [15], just to cite a few, point out a way to constrain these EoS parameters. These observations allow deviations from the Λ CDM model, which are usually parameterised by the EoS- free parameters [16–20]. In past years, there have been many observations related to the verification of the cosmic acceleration, for example, from Union 2.1[1] to the Joint LightCurve Analysis [21, 22]. But the statistics has been improved due to the density of data this kind of supernovae.

3. On how to model dark energy

One of the first steps to understand the behaviour of the cosmic acceleration remains in that we require an energy density with negative pressure at late times [23]. To achieve this, we need to express the ratio between the pressure and energy density as negative, i.e., $w(z) = P/\rho < 0$. In order to develop the evolution equations for a universe with this kind of *fluid*, we start by introducing in Einstein equations a Friedmann-Lemaitre-Robertson-Walker metric to obtain the Friedmann and Raychaudhuri equations for a spatially flat universe:

$$E(z)^2 = \left(\frac{H(z)}{H_0}\right)^2 = \frac{8\pi G}{3}(\rho_m + \rho_{DE})\left[\Omega_{0m}(1+z)^3 + \Omega_{0(DE)}f(z)\right], \tag{1}$$

and

$$\frac{\ddot{a}}{a} = -\frac{H^2}{2}[\Omega_m + \Omega_{DE}(1+3w)], \tag{2}$$

where $H(z)$ is the Hubble parameter in terms of the redshift z, G is the gravitational constant and the subindex 0 indicates the present-day values for the Hubble parameter and matter densities.

From Eq. (2), it is possible to obtain the energy conservation equation, in that way, the energy density of the non-relativistic matter is $\rho_m(z) = \rho_{0m}(1+z)^3$. And the ρ_m is given by:

$$\rho_m(z) = \rho_{0m}(1+z)^3, \tag{3}$$

and the dark energy density can be modulated as $\rho_{DE}(z) = \rho_{0(DE)}f(z)$, where can be written as:

$$\rho_{DE}(z) = \rho_{0(DE)}\, f(z). \tag{4}$$

[1] http://supernova. lbl.gov/Union/

If we assume that the energy-momentum tensor (on the right side of the Einstein's equations) $T^{\mu\nu}$ is a perfect fluid (without viscosity or stress effects), i.e., $\nabla_\mu T^{\mu\nu} = 0$, the form of $f(z)$ can be restricted to be:

$$f(z) = e^{\left[3\int_0^z \frac{1+w(\tilde{z})}{1+\tilde{z}} d\tilde{z}\right]}. \tag{5}$$

Now, the behaviour of the latter is restricted directly to the form of $w(z)$, which can give a description of the Hubble function (which can be normalised by the constant Hubble H_0), as for example, in the case of quiescence models ($w = const.$) the solution of $f(z)$ is $f(z) = (1+z)^{3(1+w)}$. If we consider the case of the cosmological constant ($w = -1$), then $f = 1$.

Some interesting insights of the above forms for $w(z)$ has been reported in [4, 24] and references therein, where a dark energy density ρ_{DE} with varying and non-varying $w(z)$ is considered.

As an extension, with the later equations we can calculate the dynamical age of the universe using the follow relationship:

$$\Omega_m + \Omega_{DE} = 1 \quad \text{or} \quad \frac{\rho_m}{\rho_{DE}} = \frac{\Omega_m}{\Omega_{DE}}. \tag{6}$$

Integrating, we can obtain:

$$t_0 = \int_0^\infty \frac{dz}{(1+z)H(z)}, \tag{7}$$

$$t_0 = H_0^{-1}\int_0^\infty \frac{dz}{(1+z)\sqrt{\left[\Omega_{0m}(1+z)^3 + \Omega_{0(DE)}f(z)\right]}}. \tag{8}$$

From here, we can set a functional form of $f(z)$, in which contribution of the dark energy density to $H(z)$ in Eq. (1) goes to a region of negative values of $w(z)$. The physics behind this behaviour is an impact on the evolution of dark energy using the dynamical age of the universe Eq. (8). When we compare several theoretical models in the light of observations, a model approach is essential. As we mentioned in the "Introduction" section, to obtain a dark energy model with late-time negative pressure, we can think in two scenarios:

- a quiescence model, which can show a wide application in tracker the slow roll condition of scalar fields and demands a constant value of w. As an example, for a flat universe and according to the Planck data [21], the dark energy EoS parameter gives $w = -1.006 \pm 0.045$, which is consistent with the cosmological constant. These data constrain the curvature parameter at 2 σ and are found to be very close to 0 with $|\Omega_k| < 0.005$.

- a kinessence model; where when the EoS is a function of redshift z. For this case, several dark energy models with different parameterisations of $w(z)$ have been discussed in the literature [24].

4. Standard dark energy models

One of the most commonly used proposals in the literature are Taylor series-like parameterisations [25–29]:

$$w(z) = \sum_{n=0} w_n x_n(z), \tag{9}$$

where w_n are constants and $x_n(z)$ are functions of the redshift z, or, the scalar factor a. As brief examples, in this section, we present three models that have bidimensional forms in the since that they depend only of two free parameters w_i. A first target is to express the exact form of the Hubble function using a specific expression for w given by Eq. (5). Once integrated, we can normalise this function by a Hubble parameter H_0, and from now on, we called this normalisation function depending of the redshift as $E(z) = H(z)/H_0$. The second target is to test these equations with the current astrophysical data available.

4.1 Lambda cold dark matter-redshift parameterisation (ΛCDM)

This model is given by:

$$E(z)^2 = \Omega_m(1+z)^3 + (1-\Omega_m), \tag{10}$$

where Ω_m represented the matter density (including the non-relativistic and dark matter terms). We consider in $f(z)$ the value of $w = -1$. As it is well known in the literature, this standard model provides a good fit for a large number of observational data surveys without addressing the important theoretical problems mentioned above.

4.2 Linear-redshift parameterisation (LR)

One of the first attempts using Taylor series—at first order—is the EoS given by [30, 31].

$$w(z) = w_0 - w_1 z, \tag{11}$$

from we can recover ΛCDM model if $w(z) = w = -1$ with $w_0 = -1$ and $w_1 = 0$. We notice immediately that due the linear term in z, this proposal diverges at high redshift and consequently yields strong constraints on w_1 in studies involving data at high redshifts, e.g., when we use CMB data [32].

As usual, we can use the later to obtain an expression for the Hubble normalised function as:

$$E(z)^2 = \Omega_m(1+z)^3 + (1-\Omega_m)(1+z)^{3(1+w_0+w_1)} \times e^{-3w_1 z} \tag{12}$$

4.3 Chevallier-Polarski-Linder parameterization (CPL)

Due the consequence of the LP parameterisation divergence, Chevallier, Polarski and Linder proposed a simple parameterisation [33, 34] that in particular can be represented by two w_i parameters that are given by a present value of the EoS w_0 and its overall time evolution w_1. The proposal is given by the expression:

$$w(z) = w_0 + \left(\frac{z}{1+z}\right) w_1, \tag{13}$$

and its evolution is

$$E(z)^2 = \Omega_m(1+z)^3 + (1-\Omega_m)(1+z)^{3(1+w_0+w_1)} \times e^{-\left(\frac{3w_1 z}{1+z}\right)}. \tag{14}$$

As we can notice, the divergence at high redshift relaxes, but still this ansatz has some problems in specific low redshift range of observations.

5. Estimating the cosmological parameters

After we have defined a specific cosmological model, we can then perform their test using astrophysical observations. The methodology can be described by a simple calculation of the usual χ^2 method and then process the MCMC chains computational runs around a certain value [observational(s) point(s)] and obtain the best fit parameter(s) of this process. Parameter estimation is usually done by computing the so-called *likelihood function* for several values of the cosmological parameters. For each data points in the parameter space, the likelihood $\mathscr{L}$ function gives the minimised probability of obtaining the observational data that was obtained if the hypothesis parameters had the given values (or priors). For example, the standard cosmological model Λ CDM is described by six parameters, which include the amount of dark matter and dark energy in the universe as well as its expansion rate H. Using the CMB data (which is the accuracy data that we understand very well so far), a likelihood function can be constructed. The information given by $\mathcal{L}$ can tell which values of these parameters are more likely, i.e. by probing many different values. Therefore, we are able to determine the values of the parameters and their uncertainties via error propagation over the free parameters of the model.

Now, the following question is that what kind of astrophysical surveys[2] can we use to test the cosmological models. In the next sections we described the most used surveys that are employed to analyse the cosmic acceleration. It is important to mention that these surveys spread depending upon their own nature. We have three types of observations classified as: standard candles (e.g., supernovae, in which characteristic function is the luminosity distance), standard rulers (e.g., supernovae, in which characteristic function is the angular/volumen distance), and the standard sirens (e.g., gravitational waves, which can be described by frequencies or chirp masses depending the observation) [35–45]. The set of all of them can describe a precise statistics, but by separate, each of them have intrinsic problems due to their physical definition. For supernovae, the luminosity distance has in their definition an integral of the cosmological model; therefore, when we perform the error propagation, the uncertainty is high. This disadvantage can be compensated by the large population of data points in the sampler. On the other hand, the uncertainty is less for standard rulers in comparison to supernovae. For this case, the definition of angular distance does not include integrals. The price that we pay in order to use this kind of sampler is that the population of data is very small (e.g., from surveys like BOSS or CMASS, we have only seven data points). Moving forward, the observation of gravitational wave standard sirens would be developed into a powerful new cosmological test due that they can play an important role in breaking parameter degeneracies formed by other observations as the ones mentioned. Therefore, gravitational wave standard sirens are of great importance for the future accurate measurement of cosmological parameters. In this part of the chapter, we are going only to develop the use of the first two kinds of observations.

---

[2] This word in the coloquial language also can be replaced by *likelihood* –do not misunderstood with the function $\mathcal{L}$. Or simple we can called as *samplers*.

6. Supernovae sampler

Along the ninety years, since their discovery, Type IA supernovae (SNIa) have been the proof of the current cosmic acceleration. The surveys have been changing given us a large population of observations, from Union 2.1[3] to the Joint LightCurve Analysis [21, 22], the data sets have been incrementing observations and also their redshift range. Currently, the Pantheon sampler, which consists of a total 1048 Type Ia supernovae (SNIa) in 40 bins [14] compressed, is the largest spectroscopically confirmed SNIa sample to date. This latter characteristic makes this sample attractive to constrain with considerably precision the free cosmological parameters of a specific model.

SNIa can give determinations of the distance modulus μ, whose theoretical prediction is related to the luminosity distance d_L according to:

$$\mu(z) = 5\log\left[\frac{d_L(z)}{1\mathrm{Mpc}}\right] + 25, \tag{15}$$

where the luminosity distance is given in units of Mpc. In the standard statistical analysis, one adds to the distance modulus the nuisance parameter μ_0, an unknown offset sum of the supernovae absolute magnitude (and other possible systematics), which is degenerate with H_0.

Now, the statistical analysis of the this sample rests on the definition of the modulus distance as:

$$\mu\left(z_j, \mu_0\right) = 5\log_{10}\left[d_L\left(z_j, \Omega_m; \boldsymbol{\theta}\right)\right] + \mu_0, \tag{16}$$

where $d_L\left(z_j, \Omega_m; \boldsymbol{\theta}\right)$ is the Hubble-free luminosity distance:

$$d_L(z, \Omega_m; \boldsymbol{\theta}) = (1+z)\int_0^z dz' \frac{1}{E(z', \Omega_m; \boldsymbol{\theta})} . \tag{17}$$

With this notation, we expose the different roles of the several cosmological parameters appearing in the equations: the matter density parameter Ω_m appears separated as it is assumed to be fixed to a prior value, while $\boldsymbol{\theta}$ is the EoS parameters w_i. These later are the parameters that we will be constraining by the data. The best fits will be obtained by minimising the quantity [46–50]:

$$\chi^2_{\mathrm{SN}}(\mu_0, \boldsymbol{\theta}) = \sum_{j=1}^{\mathcal{N}_{\mathrm{SN}}} \frac{\left(\mu\left(z_j, \Omega_m; \mu_0, \boldsymbol{\theta}\right)\right\} - \mu_{obs}\left(z_j\right))^2}{\sigma^2_{\mu,j}}, \tag{18}$$

where $\sigma^2_{\mu,j}$ are the measurement variances. And nuisance parameter μ_0 encodes the Hubble parameter and the absolute magnitude M and has to be marginalised over.

From now on, we will assume spatial flatness; therefore, the luminosity distance is related to the comoving distance D via the equation

$$d_L(z) = \frac{c}{H_0}(1+z)D(z), \tag{19}$$

where c is the speed of light, so that, using Eq. (15) we can obtain

[3] http://supernova.lbl.gov/Union/

$$D(z) = \frac{H_0}{c}(1+z)^{-1}10^{\frac{\mu(z)}{5}-5}. \tag{20}$$

The normalised Hubble function $E(z)$ can be obtained by taking the inverse of the derivative of $D(z)$ with respect to the redshift $D(z) = \int_0^z H_0 d\tilde{z}/H(\tilde{z})$. An usual alternative, instead of using the full set of parameters for this sampler, is to use the Pantheon plugin for CosmoMC to constrains cosmological models (something similar as in the case of Joint Light Curve Analysis sampler [22]).

Since we are taking nuisance parameter M in the sample, we choose the respective values of μ_0 from a statistical analysis of the Λ CDM model with Pantheon sample obtained by fixing H_0 to the Planck value given in [51]. It is common to perform this kind of fit using computational tools that can run a standard MCMC chains. In cosmology—at least at the moment this text is writing—several codes have been implemented in order to perform the statistical fit of this parameter. The lector can explore the tool called MontePython code[4] and run a standard MCMC for M using the model of their preference. As an example, if we run a Λ CDM model with this supernovae sample, the mean value obtained will be $\mu_0 = -19.63$.

7. Baryon acoustic oscillation sampler

As a standard ruler, these astrophysical observations can contribute important features by comparing the data of the sound horizon today to the sound horizon at the time of recombination (extracted from the CMB anisotropy data). Usually, the baryon acoustic distances are given as a combination of the angular scale and the redshift separation.

To define these quantities we require a relationship via the ratio:

$$d_z \equiv \frac{r_s(z_d)}{D_V(z)}, \quad \text{with} \quad r_s(z_d) = \frac{c}{H_0}\int_{z_d}^{\infty} \frac{c_s(z)}{E(z)} dz \tag{21}$$

where $r_s(z_d)$ is the comoving sound horizon at the baryon dragging epoch,

$$r_s(z_d) = \frac{c}{H_0}\int_{z_d}^{\infty} \frac{c_s(z)}{E(z)} dz \ , \tag{22}$$

and z_d is the drag epoch redshift with $c_s^2 = c^2/3\left[1 + (3\Omega_{b0}/4\Omega_{\gamma 0})(1+z)^{-1}\right]$ as the sound speed with Ω_{b0} and $\Omega_{\gamma 0}$, which are the present values of baryon and photon parameters, respectively.

We define the dilation scale as:

$$D_V(z, \Omega_m; w_0, w_1) = \left[(1+z)^2 D_A^2 \frac{cz}{H(z, \Omega_m; w_0, w_1)}\right]^{1/3}, \tag{23}$$

where D_A is the angular diameter distance given by

$$D_A(z, \Omega_m; w_0, w_1) = \frac{1}{1+z}\int_0^z \frac{c\, d\tilde{z}}{H(\tilde{z}, \Omega_m; w_0, w_1)}. \tag{24}$$

---

[4] https://monte-python.readthedocs.io/en/latest/

Using the comoving sound horizon, we can relate the distance ratio d_z with the expansion parameter h (defined such that $H \doteq 100h$) and the physical densities Ω_m and Ω_b. Therefore, we have

$$r_s(z_d) = 153.5\left(\frac{\Omega_b h^2}{0.02273}\right)^{-0.134}\left(\frac{\Omega_m h^2}{0.1326}\right)^{-0.255} \text{Mpc}, \tag{25}$$

with $\Omega_m = 0.295 \pm 0.304$ and $\Omega_b = 0.045 \pm 0.00054$ [22]. As we mentioned above, unfortunately, so far we have a very low data population of this sampler. Moreover, as an example for this text, we employed compilations of three current surveys: $d_z(z = 0.106) = 0.336 \pm 0.015$ from six-degree Field Galaxy Survey (6dFGS) [52], $d_z(z = 0.35) = 0.1126 \pm 0.0022$ from Sloan Digital Sky Survey (SDSS) [53] and $d_z(z = 0.57) = 0.0726 \pm 0.0007$ from Baryon Oscillation Spectroscopic Survey (BOSS) with high-redshift CMASS [54].

We can also, add to the full sample three correlated measurements of $d_z(z = 0.44) = 0.073$, $d_z(z = 0.6) = 0.0726$ and $d_z(z = 0.73) = 0.0592$ from the WiggleZ survey [55], which has the inverse covariance matrix:

$$\mathbf{C}^{-1}_{\textbf{WiggleZ}} = \begin{pmatrix} 1040.3 & -807.5 & 336.8 \\ -807.5 & 3720.3 & -1551.9 \\ 336.8 & -1551.9 & 2914.9 \end{pmatrix} \tag{26}$$

In order to perform the χ^2-statistic, we define the proper χ^2 function for the BAO data as

$$\chi^2_{\text{BAO}}(\boldsymbol{\theta}) = \mathbf{X}^T_{\textbf{BAO}}\mathbf{C}^{-1}_{\textbf{BAO}}\mathbf{X}_{\textbf{BAO}} \tag{27}$$

where $\mathbf{X}_{\textbf{BAO}}$ is given as

$$\mathbf{X}_{\textbf{BAO}} = \left(\frac{r_s(z_d)}{D_V(z, \Omega_m; w_0, w_1)}) - d_z(z)\right) \tag{28}$$

Then, the total χ^2_{BAO} is directly obtained by the sum of the individual quantity by using Eq. (27) in

$$\chi^2_{\text{BAO-total}} = \chi^2_{\text{6dFGS}} + \chi^2_{\text{SDSS}} + \chi^2_{\text{BOSSCMASS}} + \chi^2_{\text{WiggleZ.}} \tag{29}$$

8. How to deal with Bayesian statistics

Now, we are ready to introduce how to extrapolate the above frequentist analyses to the Bayesian field [56]. The important difference between both statistics is that in the first one we are dedicated in work with a standard χ^2 fit, while in the second one, we are taking into account the following idea: given a specific set of cosmological values (the priors), which are the probability of a second set of values to fit the hypothesis [57–60].

The above idea is what we call a Bayesian model selection, which methodology consist in describe the relationship between the cosmological model, the astrophysical data and the prior information about the free parameters. Using Bayes theorem [61], we can update the prior model probability to the posterior model probability. However, when we compare models, the evidence function is used to evaluate the model's evolution using the data at hand.

We define the evidence function as:

$$\mathcal{E} = \int \mathcal{L}(\theta)P(\theta)d\theta, \tag{30}$$

where θ is the vector of free parameters (which for the dark energy models presented in the above sections, will be given by the w_i free parameters). $P(\theta)$ is the prior distribution of these parameters.

From a computational point of view, and due to the large population of data and the model used, Eq. (30) can be difficult to calculate due that the integrations can consume to much computational time when the parametric phase space is large. Nevertheless, even when several methods exist [62, 63], in this text, we present test with a nested sampling algorithm [64] which has proven practicable in cosmology applications [65].

Once we obtain the evidence, we can therefore calculated the logarithm of the Bayes factor between two models $\mathcal{B}_{ij} = \mathcal{E}_i/\mathcal{E}_j$, where the reference model ($\mathcal{E}_i$) with highest evidence can be the Λ CDM model and impose a flat prior on H_0, i.e., we can use an exactly value of this parameter.

The interpretation of the results of this ratio can be described by a scale known as Jeffreys's scale [66], which easily can be explained as follows:

- if $\ln B_{ij} < 1$, there is no significant preference for the model with the highest evidence;
- if $1 < \ln B_{ij} < 2.5$, the preference is substantial;
- and, if $2.5 < \ln B_{ij} < 5$, it is strong; if $\ln B_{ij} > 5$, it is decisive.

9. About deep learning in cosmology

Although Bayesian evidence remains the preferred method compared with information criterions and Gaussian processes on the literature, a complete Bayesian inference for model selection—this to have a scenario where we can discriminate a pivot model from a hypothesis—is very computationally expensive and often suffers from multi-modal posteriors and parameter degeneracies. As we pointed out in the later section, the calculation of the evidence leads to large time consumption to obtain the final result.

As the study of the Large Scale Structure (LSS) of the universe indicates, all our knowledge relies on state-of-the art cosmological simulations to address a number of questions by constraining the cosmological parameters at hand using Bayesian techniques. Moreover, due to the computational complexity of these simulations, some studies look remains computationally infeasible for the foreseeable future. It is at this point where computational techniques as machine learning can have a number of important uses, even for trying to understand our universe.

The idea behind the machine learning is based on considering a neural network with a complex combination of neurons organised in nested layers. Each of these neuron implements a function that is parameterised by a set of weights W. And every layer of a neural network thus transforms one input vector—or tensor depending the dimension—to another through a differentiable function. Theoretically, given a neuron n, it will receive an input vector and the choice of an activation function A_n, the output of the neuron can be computed as:

$$h^{<t>} = A_n(h^{<t-1>} \cdot W_h + x^{<t>} \cdot W_x + b_a), \tag{31}$$

$$y^t = A_n(h^t \cdot W_y + b_y), \tag{32}$$

where $h^{<t>}$ is called the hidden state, A_n is the activation function, and y^t is the output.

The goal is to introduce a set of data in order to train this array, and therefore, the architecture can learn to finally give an output set of data. For example, the network can learn the distribution of the distance moduli in the dark energy models, then feed the astrophysical samplers (surveys) to the network to reconstruct the dark energy model and then discriminate the most probable model. [5]

Moreover, while neural networks can learn complex nested representations of the data, allowing them to achieve impressive performance results, it also limits our understanding of the model learned by the network itself. The choice of an architecture [67] can have an important influence on the performance of the neural network. Some designs have to made concerning the number and the type of layers, as well as the number and the size of the filters used in each layer. A convenient way to select these choices is typically through experimentation—which for our universe, we will need these to happen first—as it is, we can select the size of the network, which depends on the number of training test as networks with a large number of cosmological parameters likely to overfit if not enough training tests are available.

At the moment these lines are writing, a strong interest over this kind of algorithm is not only bringing new opportunities for data-driven cosmological discovery but will also present new challenges for adopting machine learning—or, in our case, a subset of this field, deep learning—methodologies and understanding the results when the data are too complex for traditional model development and fitting with statistics. A few proposals in this area have been done to explode the deep learning methods for measurements of cosmological parameters from density fields [68] and for future large-scale photometric surveys [69].

10. Deep learning for dark energy

The first target in order to start training an astrophysical survey is to design an architecture with an objective function of neural networks that can have many unstable points and local minima. This architecture makes the optimisation process very difficult, but in real scenarios [70, 71], high levels of noise degrade the training data and typically result in optimisation scenarios with more local minima and therefore increase the difficulty in training the neural network. It can thus be desirable to start optimising the neural network using noise-free data which typically yield smoother scenarios. As an example, in **Figure 1**, we present a standard network using an image of a cosmological simulation (the data) and then divided an array of several layers to finally extract the output cosmological parameters value [72, 73]. Each neuron use a Bayesian process to compute the error propagation as it is done in the standard inference analyses.

We can describe a quickly, but effective, recipe to develop a Recurrent Neural Network with a Bayesian computation training [29, 74–78] in the following steps:

---

[5] In this text we are employing a Recurrent Neural Network. There are several in this machine learning field e.g. in [57] and references therein.

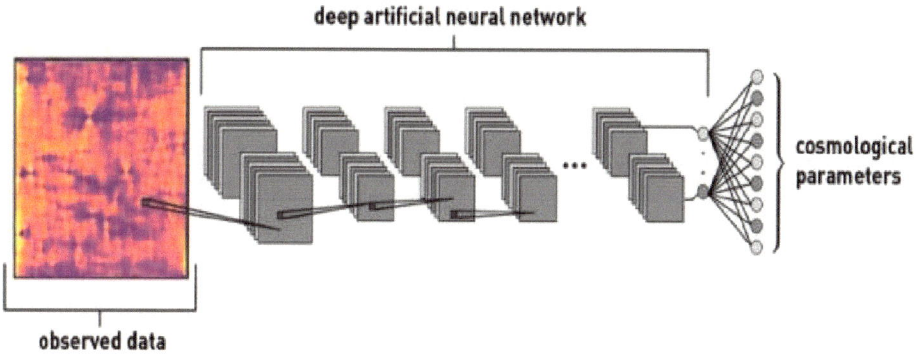

Figure 1.
A deep learning architecture for dark energy.

- Step 1. Construction of the neural network. For a Recurrent Neural Network method, we can choose values that have one layer and a certain number of neurons (e.g., you can start with 100 for a supernovae sampler).

- Step 2. Organising the data. We need to sort the sampler from lower to higher redshift in the observations. Afterwards, we re-arrange our data using the number of steps (e.g., try with four steps numbered as x_i for a supernovae sampler).

- Step 3. Computing the Bayesian training. Due to the easiness of neural networks to overfit, it is important to choose a mode of regularisation. With a Bayesian standard method to compute the evidence, the algorithm can calculate errors via regularisation methods [74]. Finally, over the cost function we can use Adam optimiser.

- Step 4. Training the entire architecture. It is suitable to consider a high number of epochs (e.g., for a sampler as Pantheon, you can try with 1000 epoch per layer). After the training, it is necessarily to read the model and apply more times the same dropout to the initial model. The result of this step is the construction of the confidence regions.

- Step 5. Computing modulus distance $\mu(z)$ for each cosmological model. Using the definitions of $E(z)$, we can compute $\mu(z)$ by using a specific dark energy equation of state in terms of z and then integrating them.

- Step 6. Computing the best fits. Finally, the output values can be obtained by using the training data as a simulated sample. We use the publicly codes CLASS[6] and Monte Python[7] to constrain the models as it is standard for usual Bayesian cosmology.

- The results of this recipe can be seeing in **Figure 2**.

[6] https://github.com/lesgourg/class_public
[7] https://github.com/baudren/montepython_public

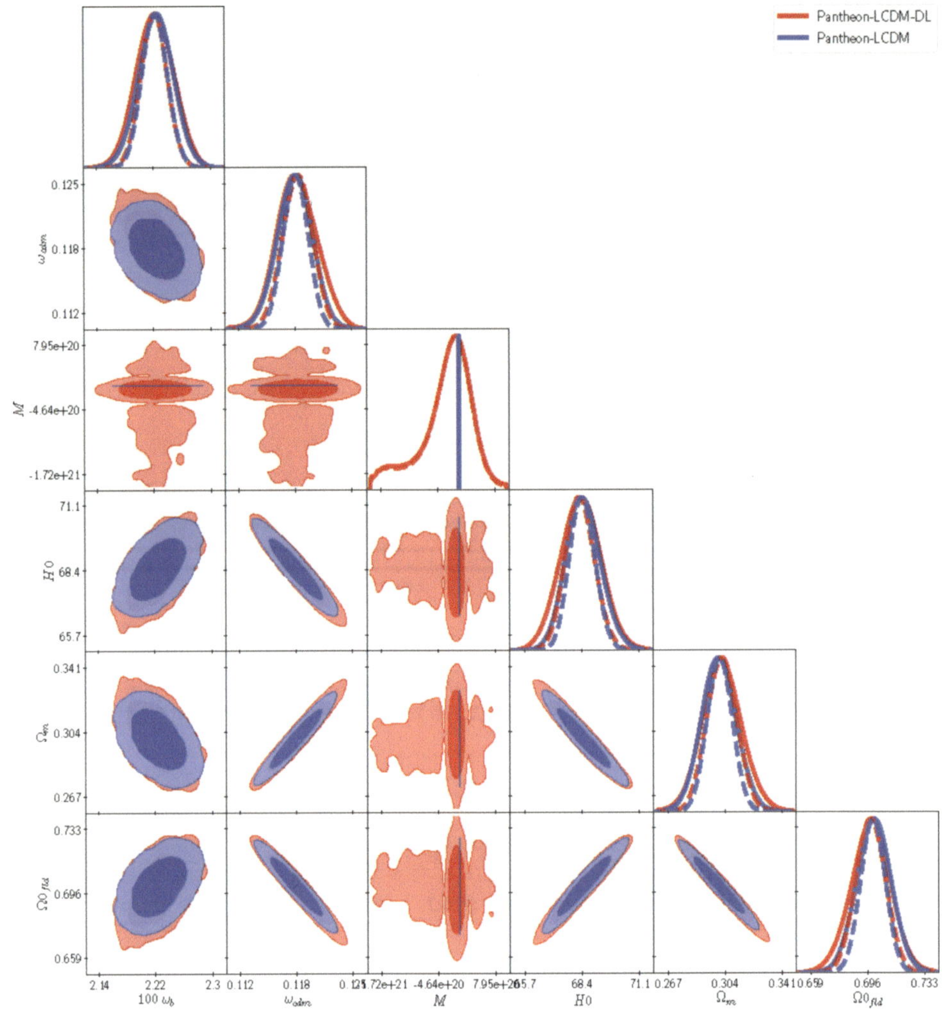

Figure 2.
Statistical contours levels for Λ CDM using observational data (red colour) and training deep learning data (blue colour).

11. Conclusions

In this chapter, we discuss how to derive the equations of state for a specific dark energy model. Also, we studied the standard models of dark energy in order to project the cosmic acceleration according to the current data available in the literature. It is important to remark that each Bayesian statistics performed will depend solely on the data used to develop them. More the data, better the statistics. So we expect that future surveys will improve the constrains over the cosmological parameters, not only at background level, but also at perturbative level.

The exploration of these astrophysical surveys has reached a new scenario in regards to the machine learning techniques. These kind of techniques allow to explore—without technical problems in the astrophysical devices—scenarios where the pivot model of cosmology, ΛCDM, a theoretical framework that accurately describes a large variety of cosmological observables, from the temperature anisotropies of the cosmic microwave background to the spatial distribution of galaxies. This model has a few free parameters representing fundament quantities, like the geometry and expansion rate of the Universe, the amount and nature of dark energy, and the sum of neutrino masses. Knowing the value of these parameters will improve our knowledge on the fundamental constituents and laws governing our universe. Thus, one of most important goals of modern cosmology is to constrain the value of these parameters with the highest accuracy. Therefore, as an extrapolation between the ideas of the standard cosmostatistics and the use of machine learning techniques will improve even better the constrain of the cosmological parameters without to be worried about the intrinsic uncertainties of the data [79].

Acknowledgements

CE-R is supported by the *Royal Astronomical Society* as FRAS 10147, *PAPIIT* Project IA100220 and ICN-UNAM projects.

Author details

Celia Escamilla-Rivera
Instituto de Ciencias Nucleares, Universidad Nacional Autónoma de México, México D.F., México

*Address all correspondence to: celia.escamilla@nucleares.unam.mx

References

[1] Weinberg S. The Cosmological Constant Problems. 2000, arXiv:astro-ph/0005265

[2] Sahni V, Starobinsky AA. The case for a positive cosmological lambda term. International Journal of Modern Physics D: Gravitation; Astrophysics and Cosmology. 2000;**9**:373-443

[3] Feng L, Lu T. A new equation of state for dark energy model. Journal of Cosmology and Astroparticle Physics. 2011;**2011**:034

[4] Stefancic H. Equation of state description of the dark energy transition between quintessence and phantom regimes. Journal of Physics Conference Series. 2006;**39**:182

[5] Wang Y, Tegmark M. Uncorrelated measurements of the cosmic expansion history and dark energy from supernovae. Physical Review D. 2005; **71**:103513

[6] Barboza EM, Alcaniz JS, Zhu Z-H, Silva R. A generalized equation of state for dark energy. Physical Review D. 2009;**80**:043521

[7] Pantazis G, Nesseris S, Perivolaropoulos L. Comparison of thawing and freezing dark energy parametrizations. Physical Review D. 2016;**93**:103503

[8] Jassal HK, Bagla JS, Padmanabhan T. WMAP constraints on low redshift evolution of dark energy. Monthly Notices of the Royal Astronomical Society. 2005;**356**:L11

[9] Wang Y. Physical Review D. 2008;**77**: 123525. DOI: 10.1103/PhysRevD. 77.123525 [arXiv:0803.4295 [astro-ph]]

[10] Escamilla-Rivera C, Capozziello S. Unveiling cosmography from the dark energy equation of state. International Journal of Modern Physics D. 2019. DOI: 10.1142/S021827181950 1542 [arXiv: 1905.04602 [gr-qc]]

[11] Jaime LG, Patiño L, Salgado M. Note on the equation of state of geometric dark-energy in f(R) gravity. Physical Review D. 2014;**89**(8):084010. DOI: 10.1103/PhysRevD.89.084010 [arXiv: 1312.5428 [gr-qc]]

[12] Lazkoz R, Ortiz-Baños M, Salzano V. f(R) gravity modifications: From the action to the data. European Physical Journal C. 2018;**78**(3):213. DOI: 10.1140/ epjc/s10052-018-5711-6 [arXiv: 1803.05638 [astro-ph.CO]]

[13] Capozziello S, D'Agostino R, Luongo O. International Journal of Modern Physics D: Gravitation; Astrophysics and Cosmology. Extended Gravity Cosmography. 2019;**28**(10): 1930016. DOI: 10.1142/ S0218271819300167 [arXiv:1904.01427 [gr-qc]]

[14] Scolnic DM et al. The Astrophysical Journal. 2018;**859**:101

[15] Busca NG, Delubac T, Rich J, Bailey S, Font-Ribera A, Kirkby D, et al. Baryon acoustic oscillations in the Ly- α forest of BOSS quasars. Astronomy and Astrophysics. 2013;**552**:A96

[16] Alberto Vazquez J, Bridges M, Hobson MP, Lasenby AN. Reconstruction of the dark energy equation of state. JCAP. 2012;**1209**:020. DOI: 10.1088/1475-7516/2012/09/020 [arXiv:1205.0847 [astro-ph.CO]]

[17] Seikel M, Clarkson C, Smith M. Reconstruction of dark energy and expansion dynamics using Gaussian processes. JCAP. 2012;**06**:036. DOI: 10.1088/1475-7516/2012/06/036 [arXiv: 1204.2832]

[18] Montiel A, Lazkoz R, Sendra I, Escamilla-Rivera C, Salzano V.

Nonparametric reconstruction of the cosmic expansion with local regression smoothing and simulation extrapolation. Physical Review D. 2014;**89**(4):043007. DOI: 10.1103/PhysRevD.89.043007 [arXiv:1401.4188 [astro-ph.CO]]

[19] Zhao GB et al. Dynamical dark energy in light of the latest observations. Nature Astronomy. 2017;**1**(9):627. DOI: 10.1038/s41550-017-0216-z [arXiv: 1701.08165 [astro-ph.CO]]

[20] Jaime LG, Jaber M, Escamilla-Rivera C. Physical Review D. 2018;**98**(8): 083530. DOI: 10.1103/ PhysRevD.98.083530 [arXiv: 1804.04284 [astro-ph.CO]]

[21] Ade PAR, Aghanim N, Arnaud M, Ashdown M, Aumont J, Baccigalupi C, et al. Planck 2015 Results. XIII. Cosmological Parameters. 2015, arXiv: astro-ph.CO/1502.01589

[22] Betoule M et al. [SDSS Collaboration]Astronomy and Astrophysics. 2014;**568**:A22. DOI: 10.1051/0004-6361/201423413 [arXiv: 1401.4064 [astro-ph.CO]]

[23] Huterer D, Turner MS. Probing the dark energy: Methods and strategies. Physical Review D. 2001;**64**:123527

[24] Lazkoz R, Nesseris S, Perivolaropoulos L. Exploring cosmological expansion Parametrizations with the gold SnIa dataset. Journal of Cosmology and Astroparticle Physics. 2005;**2005**:010

[25] Barboza EM Jr, Alcaniz JS. A parametric model for dark energy. Physics Letters B. 2008;**666**:415-419

[26] Wetterich C. Phenomenological parameterization of quintessence. Physics Letters B. 2004;**594**:17-22

[27] Wetterich C. Cosmology with Varying Scales and Couplings. 2003, arXiv:hep-ph/0302116

[28] Escamilla-Rivera C, Casarini L, Fabris JC, Alcaniz JS. Linear and non-linear perturbations in dark energy models. 2016, arXiv:1605.01475

[29] Escamilla-Rivera C, Fabris JC. Galaxies MPDI. Galaxies. 2016;**4**(4):76. DOI: 10.3390/galaxies4040076 [arXiv: 1511.07066 [astro-ph.CO]]

[30] Weller J, Albrecht A. Future supernovae observations as a probe of dark energy. Physical Review D. 2002; **65**:103512

[31] Huterer D, Turner MS. Physical Review D. 2001;**64**:123527. DOI: 10.1103/PhysRevD.64.123527 [astro-ph/0012510]

[32] Wang FY, Dai ZG. Constraining dark energy and cosmological transition redshift with type Ia supernovae. Chinese Journal of Astronomy and Astrophysics. 2006;**6**:561

[33] Linder EV. The dynamics of quintessence, the quintessence of dynamics. General Relativity and Gravitation. 2008;**40**:329-356

[34] Chevallier M, Polarski D. Accelerating universes with scaling dark matter. International Journal of Modern Physics D: Gravitation; Astrophysics and Cosmology. 2001;**10**:213-223

[35] Albrecht A, Amendola L, Bernstein G, Clowe D, Eisenstein D, Guzzo L, et al. Findings of the Joint Dark Energy Mission Figure of Merit Science Working Group. 2009, arXiv:0901.0721

[36] Liddle AR. How many cosmological parameters? Monthly Notices of the Royal Astronomical Society. 2004;**351**: L49-L53

[37] Riess AG et al. Observational evidence from supernovae for an accelerating universe and a cosmological constant. [Supernova Search Team] Astronomy Journal. 1998;**116**:1009.

DOI: 10.1086/300499 [astro-ph/9805201]

[38] Perlmutter S et al. Measurements of Omega and Lambda from 42 high-redshift supernovae. [Supernova Cosmology Project Collaboration]The Astrophysical Journal. 1999;**517**:565. DOI: 10.1086/307221 [astro-ph/9812133]

[39] Available from: http://desi.lbl.gov/

[40] Available from: https://www.darkenergysurvey.org/

[41] Available from: https://www.lsst.org/

[42] Available from: https://wfirst.gsfc.nasa.gov/

[43] Takada M, Jain B. The three-point correlation function in cosmology. Monthly Notices of the Royal Astronomical Society. 2003;**340**:580. DOI: 10.1046/j.1365-8711.2003.06321.x [astro-ph/0209167]

[44] Marin FA et al. The WiggleZ Dark energy survey: Constraining galaxy bias and cosmic growth with 3-point correlation functions. [WiggleZ Collaboration] Monthly Notices of the Royal Astronomical Society. 2013;**432**: 2654. DOI: 10.1093/mnras/stt520 [arXiv:1303.6644 [astro-ph.CO]]

[45] Tsujikawa S. Dark energy: Investigation and modeling. 2010. DOI: 10.1007/978-90-481-8685-3_8, arXiv: 1004.1493 [astro-ph.CO]

[46] Press WH, Teukolsky A, Vetterling W, Flannery B. Numerical Recipes. 3rd ed. New York, USA: Cambridge Press; 1994

[47] Escamilla-Rivera C, Lazkoz R, Salzano V, Sendra I. Tension between SN and BAO: Current status and future forecasts. Journal of Cosmology and Astroparticle Physics. 2011. DOI: 10.1088/1475-7516/2011/09/003

[48] Burigana C, Destri C, de Vega HJ, Gruppuso A, Mandolesi N, Natoli P, et al. Forecast for the Planck precision on the tensor to scalar ratio and other cosmological parameters. The Astrophysical Journal. 2010;**724**:588

[49] Bull P et al. Physics in the Dark Universe. 2016;**12, 56**. DOI: 10.1016/j.dark.2016.02.001 [arXiv:1512.05356 [astro-ph.CO]]

[50] Sendra I, Lazkoz R. SN and BAO constraints on (new) polynomial dark energy parametrizations: Current results and forecasts. Monthly Notices of the Royal Astronomical Society. 2012; **422**:776. DOI: 10.1111/j.1365-2966.2012.20661.x [arXiv:1105.4943 [astro-ph.CO]]

[51] Aghanim N, et al. [Planck Collaboration], arXiv:1807.06209 [astro-ph.CO].

[52] Beutler F, Blake C, Colless M, Jones DH, Staveley-Smith L, Campbell L, et al. The 6dF galaxy survey: Baryon acoustic oscillations and the local Hubble constant. Monthly Notices of the Royal Astronomical Society. 2011;**416**:3017-3032

[53] Anderson L, Aubourg É, Bailey S, Beutler F, Bhardwaj V, Blanton M, et al. The clustering of galaxies in the SDSS-III baryon oscillation spectroscopic survey: Baryon acoustic oscillations in the data releases 10 and 11 galaxy samples. Monthly Notices of the Royal Astronomical Society. 2014;**441**:24-62

[54] Xu X, Padmanabhan N, Eisenstein DJ, Mehta KT, Cuesta AJ. A 2% distance to z = 0.35 by reconstructing baryon acoustic oscillations–II: Fitting techniques. Monthly Notices of the Royal Astronomical Society. 2012;**427**:2146-2167

[55] Blake C, Brough S, Colless M, Contreras C, Couch W, Croom S, et al.

The WiggleZ dark energy survey: Joint measurements of the expansion and growth history at z ¡ 1. Monthly Notices of the Royal Astronomical Society. 2012; **425**:405-414

[56] Escamilla-Rivera C, Lazkoz R, Salzano V, Sendra I. JCAP. 2011;**1109**: 003. DOI: 10.1088/1475-7516/2011/09/003 [arXiv:1103.2386 [astro-ph.CO]]

[57] Verde L,Treu T, Riess AG, arXiv: 1907.10625 [astro-ph.CO]

[58] Ratra B, Peebles PJE. Cosmological consequences of a rolling homogeneous scalar field. Physical Review D. 1988;**37**: 3406. DOI: 10.1103/PhysRevD.37.3406

[59] Armendariz-Picon C, Mukhanov VF, Steinhardt PJ. Physical Review Letters. 2000;**85**:4438. DOI: 10.1103/PhysRevLett.85.4438 [astro-ph/0004134]

[60] Escamilla-Rivera C. Status on bidimensional dark energy parameterizations using SNe Ia JLA and BAO datasets. Galaxies. 2016;**4**(3):8. DOI: 10.3390/galaxies4030008 [arXiv: 1605.02702 [astro-ph.CO]]

[61] Bayes RT. An essay toward solving a problem in the doctrine of chances. Philosophical Transactions. Royal Society of London. 1764;**53**:370-418

[62] Gregory P. Bayesian Logical Data Analysis for the Physical Sciences. New York, USA: Cambridge University Press; 2005

[63] Trotta R. Applications of Bayesian model selection to cosmological parameters. Monthly Notices of the Royal Astronomical Society. 2007;**378**: 72-82

[64] Skilling J. Bayesian Annal. 2006:833. Available from: http://www.mrao.cam.ac.uk/steve/maxent2009/images/skilling.pdf

[65] Liddle AR, Mukherjee P, Parkinson D, Wang Y. Present and future evidence for evolving dark energy. Physical Review D. 2006;**74**:123506

[66] Jeffreys H. Theory of Probability. 3rd ed. Oxford, United Kingdom: Oxford University Press; 1998

[67] Ntampaka M, et al. arXiv: 1902.10159 [astro-ph.IM]

[68] Schmelzle J, Lucchi A, Kacprzak T, Amara A, Sgier R, Réfrégier A, et al. arXiv:1707.05167 [astro-ph.CO]

[69] Charnock T, Moss A. The Astrophysical Journal. 2017;**837**(2):L28. DOI: 10.3847/2041-8213/aa603d [arXiv: 1606.07442 [astro-ph.IM]]

[70] Moss A, arXiv:1810.06441 [astro-ph.IM]

[71] Moss A, arXiv:1903.10860 [astro-ph.CO]

[72] Aurelien G. Hands-On Machine Learning with Scikit-Learn and Tensorflow: Concepts, Tools, and Techniques to Build Intelligent Systems. O'Reilly Media; 2017. https://www.oreilly.com/conferences/

[73] Kessler R, Conley A, Jha S, Kuhlmann S, arXiv:1001.5210 [astro-ph.IM]

[74] Géron A. Hands-On Machine Learning with Scikit-Learn & TensorFlow. O'REILLY; 2017. https://www.oreilly.com/conferences/

[75] Goodfellow I, Bengio Y, Courville A. Deep Learning. USA: MIT Press; 2016. Available from: http://www.deeplearningbook.org

[76] Zaremba W, Sutskever I. arXiv: 1505.00521 [cs.LG]

[77] Pedamonti D. arXiv:1804.02763 [cs.LG]

[78] Gal Y, Ghahramani Z. A Theoretically Grounded Application of Dropout in Recurrent Neural Networks. NIPS. 2016 [arXiv:1512.05287v5]

[79] Escamilla-Rivera C, Quintero MAC, Capozziello S. A deep learning approach to cosmological dark energy models. JCAP. 2019;(3). DOI: 10.1088/1475-7516/2020/03/008. arXiv:1910.02788 [astro-ph.CO]

Chapter 3

The Tension over the Hubble-Lemaitre Constant

Michael L. Smith and Ahmet M. Öztaş

Abstract

Astronomers continue the search for better a Hubble-Lemaitre constant, H_0, value and other cosmological parameters describing our expanding Universe. One investigative school uses 'standard candles' to estimate distances correlated with galactic redshifts, which are then used to calculate H_0 and other parameters. These distance values rely on measurements of Cepheid variable stars, supernovae types Ia and II or HII galaxies/giant extra-galactic HII regions (GEHR) or the tip of red giant star branch to establish a distance 'ladder'. We describe some common pitfalls of employing log-transformed HII/GEHR and SNe Ia data rather than actual distances and suggest better analytical methods than those commonly used. We also show that results using HII and GEHR data are more meaningful when low quality data are discarded. We then test six important cosmological models using HII/GEHR data but produce no clear winner. Groups utilising gravitational waves and others measuring signals from the cosmic microwave background are now at odds, 'tension', with those using the SNe Ia and HII data over Hubble-Lemaitre constant values. We suggest a straightforward remedy for this tension.

Keywords: Hubble constant, Hubble-Lemaitre constant, cosmological parameters, distance ladder, distance scale, data analysis, supernova, HII galaxies, luminosity distance, redshift

1. Introduction

Some important goals of cosmology are determination of values for the local Hubble-Lemaitre constant, H_0, the average Universe matter density, ρ, as well as confirmation, or not, of the cosmological constant, Λ. Important tools are information emanating from supernovae types Ia (SNe Ia) and II (SNe II) explosions, γ-ray bursts and redshifts, z, combined with distance determinations to closer Cepheid variable stars [1, 2]. Estimates are also made using data from single events; the cosmic microwave background (CMB) [3, 4] and gravitational waves combined with electromagnetic detection [5, 6]. Values for H_0 as determined by different research groups do not closely match and the situation is described as 'tension'; the differences being ascribed to the lack of dark energy and the larger matter density in our early Universe [7] or perhaps to a 'local hole' and discrepancies between parallax distance estimates [8]. The different values for H_0 are puzzling for one might expect larger values towards recombination than today but the reverse is reported and the difference between H_0 values is increasing with more reports [9]. The controversy has generated considerable interest for astrophysics including Internet blogs and one video with over 135,000 views [10].

Results from the CMB investigations depend on exacting measurements of tiny, low-energy fluctuations modelled with at least 6 parameters, demanding many 'priors' (fixed-valued parameters) and cannot realistically discriminate between models since there are many parameter combinations able to fit many models [3, 11]. Results from SNe Ia investigations are model-dependent, rely on 2 or 3 parameters as published, but in reality 4 or 5 para-meters are used for modelling and the belief that most SNe Ia events are uniform and similar. Systematic errors of collection and analysis are still being discovered, corrected or culled from data [12].

Estimates of SNe Ia distances typically rely on nearby Cepheid variable star distances which are still being adjusted [13, 14]. In addition, the methods used for evaluations of the SNe Ia data and claims therefrom have been repeatedly questioned [15, 16]. An independent method for estimating H_0 has recently been published based on the characteristics of selected red giant stars [17, 18]. The value found, 69.8 km s^{-1} Mpc^{-1} is close to the gravitational wave observation from a bi-neutron star collision, 70 km s^{-1} Mpc^{-1}, much lower than calculated with SNe Ia data [1].

A pioneering effort is being made using the L(Hβ)σ distance estimator for giant extra-galactic HII regions (GEHR) and HII galaxies back to $z \approx 2.3$ by a small group [19–23]; significantly further than SNe Ia observations. Several assumptions and adjustments to the data are necessary to allow use as astronomical distances analogous to SNe Ia data as done by Wei et al. [24]. The latter group presented results using the HII distance magnitudes, *mag*, and redshifts collected by the former group to investigate the properties of three important cosmological models. Their results suggest the Rh = ct model is a slightly better fit to the HII and GEHR data than two, well-known versions of the current standard (ΛCDM, ωΛCDM) models [25, 26]. The Rh = ct model is a special, geometrically flat version of an earlier proposal by John and Joseph [27]. This eternal, coasting, non-empty model has been recently reviewed [28]. Analyses using HII and GEHR data with the Rh = ct model suggest a H_0 of 62.3 km s^{-1} Mpc^{-1} [24, 29], which is lower than other recent reports but exactly the last estimate of Sandage and coworkers based on data from the Hubble telescope [30]. This low value is very important if true, because it greatly increases the Universe age allowing more time for initial star and galaxy formations. A second, recent study claims a selection of the HII data supports a flat Universe geometry and the Rh = ct model [31].

A major problem with these analyses is the use of a relationship commonly termed as a *Hubble diagram* or a *Hubble relationship* but is not. This relationship and analytic technique are commonly used in astrophysics and often leads to erroneous results. The real Hubble relationship is based on correlation between distance and galaxy recession velocity in accordance with fundamental physics [32]. Unfortunately, it is now common to correlate some version of log(distance), usually termed *distance magnitude* or simply *mag*, with redshift (*mag* versus z, sometimes displayed as log[z]), and still use the term Hubble diagram to describe this correlation, which certainly is not distance versus velocity. Here we use the term *pseudo-H-routine* to describe the use of *distance mag* versus z and *H-routine* for luminosity distance, D_L versus relative recession velocity, ν/ν_0 or a, the expansion factor.

There are many drawbacks using the pseudo-H-routine for model testing. First, distance is a physical metric but *mag* is not. Second, this routine non-uniformly compresses data dispersion and standard errors; errors of distant observations are systematically compressed over errors of more nearby emissions and will exacerbate skewness [33]. Using weighed regression analysis the pseudo-H-routine incorrectly emphasises the more imprecise, distant data, SNe or HII, which often leads to incorrect regression minima and results [34, 35]. Third, the best data pair, recession velocity and position of earth or the local group (1,0) without error, cannot be used with the pseudo-H-routine, the distance becoming $-\infty$; this exclusion can never be justified. Fourth, because the errors have been compressed, goodness of fit estimates are not properly distributed, are both smaller with more similar values, complicating model discrimination. Fifth, information from both intercepts are lost and cannot be recovered. If the pseudo-H-routine were valid, parameter estimates should be similar between the Hubble relationship and the pseudo-H-routine, but are not. Here and for other examples using SNe Ia data, the two analytic methods do not agree [16, 36].

There are many reasons to perform regression analysis using luminosity distance versus expansion factor (D_L versus a, H-routine) rather than *mag* versus z. First, the real distances and errors are used rather than perturbations. This allows examination of the real data dispersion and proper estimation of regression best fit. Second, more realistic measurement errors of distant objects are used for H-routine and are not dampened. The best error estimates are required to properly employ weighed fitting for regression routines. Third, the very best data pair (1,0) without error, is automatically used and anchors the regression estimates. For cosmology the position and velocity of the earth, sun or local group, are the very best anchor and estimates of the origin can be estimated directly using the H-routine. Fourth, the H-routine comes with two intercepts, the position of the earth and the Universe age. With good data the latter may be guesstimated, being asymptotic, and the former may be used to judge the value of nearby distances. Fifth, the H-routine allows visual examination of the relative data worth, which is important when these are billions of years old. Some SNe Ia and HII emissions display standard deviations similar to the Universe age. Sixth, the difference between goodness of fit estimates, such as χ^2, are not depressed, which eases objective model discrimination. We claim the famous presentation of Hubble is still valid and suggest analyses using the pseudo-H-routine should be avoided.

Here we first use the same routine as Wei et al. (*mag* versus z) to check our methods. Next, to comply with the requirements of the H-routine, we calculate the luminosity distances, D_L, and use the related expansion factors to test six cosmological models. We begin using all 156 HII and GEHR observations [20, 22, 24] but find it necessary to parse the data to obtain worthwhile results so we are forced to discard about half the events even though we conservatively allow all data within 99.99% confidence or a Studentised limit of 1.5. After this we find our analyses can more easily discriminate the relative model worth. Our results only support the precedence of the Rh = ct model when all HII/GEHR data are used. When questionable data are discarded, the $\omega\Lambda$CDM model is a better fit to the HII/GEHR data, however, this model requires 4 free parameters and in reality 6 parameters which means it is a plastic model which overfits the data. We suggest more and better HII and GEHR data are needed before this independent tool can be confidently used for model discrimination.

2. Data, models and methods

2.1 HII and GEHR data

We use data; *mag*, standard deviations about *mag* and redshifts, z, adjusted to the local group rather than heliocentric, as tabulated by Wei et al. [24] with HII data from [20, 23] and data collected and analysed by [22]. The HII data are usually those with $a < 0.85$ and the GEHR data are usually associated with $a > 0.85$. Not all values from [20] were used by Wei leaving a total of 156 HII/GEHR data pairs. For examination using the relationship of distance modulus, *mag* versus z the pseudo-H-routine, we follow [24] with H_0, nuisance parameter and *mag* value adjustments during iterative regression calculations. For model examination using the H-routine, we calculate the actual distances, D_L in Mpc, with the usual relationship, Eq. (6) of [24] as

$$D_L = 10((mag - 25)/5) \tag{1}$$

where *mag* is the distance modulus. We perform H_0 and D_L value and standard deviation re-adjustments during repetitive calculations followed by re-evaluation and recalculation of H_0, D_L and associated estimated errors, since the values of *mag* and D_L are dependent on H_0; details in [24]. The expansion factors a, strictly proportional to the recession velocities, are calculated from $a = 1/(1 + z) = \nu/\nu_0$ and are not modified, since observational errors of redshift determinations are tiny compared to errors about D_L. Though this has been recently contested if the reader carefully examines the data they will agree with our assessment of relative error [37].

For regression using D_L versus a we check the consistency of the different data sets compiled in [24] by calculating the D_L versus a from the tabulated *mag* values, as per Eq. (1) and use the position and recession velocity of the local group (1,0) as the anchor. We also use the distance and expansion factor values for 10 nearby galaxies but found these unnecessary as anchors, details in the Results. We check several different data subsets and find the different subsets to be internally consistent, that is, displaying random distribution about the best fits with the exception of 25 observations from $a < 0.86$ ($z \approx 0.17$) of [23]. When only these values are tested with the local group as the 'anchor' the distances to generally unbelievably large with very large standard deviations and inconsistent with the anchor. Rather than systematically reducing the D_L values of distant emissions we use the 156 tabulated values from [24].

To suppress the influence of outliers we parse the data in two manners. For both situations we only discard data simultaneously failing three models, ΛCDM, ωΛCDM and $Rh = ct$, at all three H_0 values of 74.3, 71.0, 62.3 km s^{-1} Mpc^{-1}. For the first parse we discard those beyond the 99.99% limits ($\approx 4\sigma$), leaving from 86 to 89 data pairs. For the second subset we discard data exhibiting $ri/\sigma i > 1.5$ (residual/standard deviation; the Studentised limit) [38], leaving 74 to 77 pairs. Even using our conservative parsing routines both methods trim a number of observations from $0.85 < a < 1$. After parsing we perform regression analyses using the H-routine.

2.2 Models

The models tested are based on the Friedmann-Lemaitre-Robertson-Walker (FLRW) universe; explanations can be found in sources [39–41]. This is by far the most useful model of cosmology and an early version was used by Einstein and de Sitter to model the Universe, subsection 2.2.4. We make the usual assumption for FLRW model parameter normalisation

$$1 = \Omega_m + \Omega_\Lambda + \Omega_k \tag{2}$$

where Ω_m, Ω_Λ, Ω_k are normalised matter density, cosmological constant (dark energy) and spacetime curvature, respectively. Four models examined here presume flat spacetime geometry with $\Omega_k = 0$, Eqs. (3), (5), (7), and (10). We will not re-examine the HII/GEHR data using Eqs. (4), (6) and (8) which have been used by others many times.

2.2.1 Eternal coasting, Rh = ct model

This is the preferred model of Melia and coworkers [28, 42] with only one free parameter, H_0, presuming a geometrically flat universe and we use the relationship

$$D_L = \frac{c}{H_0 a} \ln \frac{1}{a} \tag{3}$$

with a the expansion factor, c is lightspeed and the natural ln. Eq. (2) may hold true testing this *Rh* = ct model with HII/GEHR and SNe Ia data but Ω_m, Ω_k and Ω_Λ are not explicit with this model so cannot be evaluated, leaving H_0 as the single parameter. To use the *mag* with redshift, z, one must employ not Eq. (3) but with

$$mag = 5\log\left(\frac{c(1+z)}{H_0}\right)\ln(1+z) + 25 \tag{4}$$

Using Eq. (4) allows presentation of the pseudo-H-diagram with HII/GEHR data as in [24] where it is labelled 'Hubble diagram' though not really a Hubble-type diagram [32].

2.2.2 The current standard model of cosmology, ΛCDM

Called the standard model of cosmology by some, with two free parameters, after adjusting the data for the local value of H_0 as

$$D_L = \frac{c}{H_0 a}\int_1^a \left(\frac{da}{a\sqrt{\frac{\Omega_m}{a} + (1-\Omega_m)a^2}}\right) \tag{5}$$

where $(1 - \Omega_m)$ represents the normalised Ω_Λ. The effect of spacetime curvature, Ω_k, is presumed negligible for a geometrically flat universe. What is used in practice are values for the *distance magnitude* with the *mag* and z data modelled with this equation

$$mag = 5\log\left[\frac{c(1+z)}{H_0}\int_0^z \left(\frac{dz}{\sqrt{(1+z)^2(1+\Omega_m z) + 2(2+z)(1-\Omega_m)}}\right)\right] + 25 \tag{6}$$

but this version is rarely made explicit. Typical diagrams using this relationship can be found in [1, 43] and in award winning articles [44], but these are not really Hubble-type diagrams.

2.2.3 The standard model allowing the equation of state (EoS), ωΛCDM

This is a variant of the standard model with a parameter, ω, directly effecting Ω_Λ as in

$$D_L = \frac{c}{H_0 a}\int_1^a \left(\frac{da}{a\sqrt{\frac{\Omega_m}{a} + a^2(1-\Omega_m)^{3(1+\omega)}}}\right) \tag{7}$$

where ω is a parameter estimating the relative influences of dark energy p and ρ as (p = ωρ)de being the dark energy equation of state. A flat universe geometry is preferred with $\omega \approx -1$ being a target by many [40]. What is used in practice with the *mag* and z data is

$$mag = 5\log\left[\frac{c(1+z)}{H_0}\int_0^z\left(\frac{dz}{\sqrt{(1+z)^2(1+\Omega_m z)+2(2+z)(1-\Omega_m)^{3(1+\omega)}}}\right)\right]+25 \quad (8)$$

but this version, too, is almost never presented. The reader can see that it is possible to add an extra parameter as a simple term to Eqs. (3), (5) and (7) which may be evaluated as the intercept. Results presenting large intercept values means some form of systematic error is present in the data. It is meaningless to add an extra term to Eqs. (4), (6) and (8) since evaluation cannot be made at the origin, which is $-\infty$ so one loses a simple method for evaluating data worth.

2.2.4 The Einstein-de Sitter model, EdS

After consulting with E. Hubble and becoming a convert to the idea of a dynamic universe, Einstein reconsidered the basis for his field equation. He then dropped the cosmological constant, Λ, and solved his theory for an expanding universe consisting of only matter as reviewed in [45]. Einstein and de Sitter began with the assumptions of Friedmann demanding spacetime curvature due to the presence of matter, which we cast using the FLRW model as

$$D_L = \frac{c}{aH_0}\,sinh\left[\int_a^1\left(\frac{da}{a\sqrt{\frac{\Omega_m}{a}}}\right)\right] \quad (9)$$

where we use the *sinh* function for positive curvature. If we allow $\Omega_m = 1$ the result is a relationship which we term the EdS model. We have previously pointed out this model functions very well describing the Universe using only nearby emissions, $z < 0.10$ [46].

$$D_L = \frac{c}{aH_0}\,sinh\left(\sqrt{1-a}\right) \quad (10)$$

2.2.5 The cosmological general relativity model, CGR

The cosmological general relativity model is a recent introduction to this subject [47]. The Hubble velocity is used as a tool by CGR to aid solutions using a 5-dimensional tensor with some success. For instance, the CGR model has been successfully applied to the Tully-Fisher relationship describing spiral galaxies [48]. This model does not admit the existence of dark matter and since dark matter has not been confirmed by several sensitive, direct observational tests [49, 50] this constraint holds true, so this model deserves our consideration.

For estimating H_0 using standard candles, the CGR model for a flat universe present

$$D_L = \left(\frac{c}{aH_0\sqrt{1-\beta^2}}\right)\times\left(\frac{sinh\left(\beta\sqrt{1-\Omega_\beta}\right)}{\sqrt{1-\Omega_\beta}}\right) \quad (11)$$

where β is shorthand for

$$\beta = \frac{\left(\frac{1}{a}\right)^2 - 1}{\left(\frac{1}{a}\right)^2 + 1} \tag{12}$$

and Ω_b is the estimate of baryonic (normal) matter, whereas Ω_m applies to all matter types. The reader should remember that baryonic matter is supposedly only 15–20% as plentiful as dark matter.

2.2.6 The FLRW model, ST, with the term, Ω_k, for spacetime

This model, which we term ST, admits the influences of both matter and geometric curvature on universe expansion so the condition for normalisation is

$$1 = \Omega_m + \Omega_k. \tag{13}$$

This is a significant change from other models considered here since the ST model is not restricted to a flat universe but allows the spacetime curvature parameter, Ω_k. For use with the HII and GEHR data we present the form which can be integrated numerically as

$$D_L = \frac{c}{H_0 a} \int_1^a \left(\frac{da}{a\sqrt{\frac{\Omega_m}{a} + (1 - \Omega_m)}} \right) \tag{14}$$

or can be integrated analytically [46]. Comparison of the above with Eq. (5) shows these be similar except replacement of the $1 - \Omega_m$ term with $(1 - \Omega_m)a^2$ if one presumes a flat universe and late dark energy. This model deserves consideration since a recent attempt to detect dark energy in a lab failed [51].

2.3 Methods

We first apply the pseudo-H-routine to the log-transformed data as per Eqs. (4), (6) and (8) to check our regression routines. The reader should note that not only are the dependent values transformed into a logarithmic metric but the abscissa is transformed from relative velocity into galactic redshift as another non-linear metric, $z = \upsilon/\upsilon_0 - 1$. We use a weighted, robust regression routine normalised as

$$1 = \sum_1^i \ln\left(1 + \left(|y_i - \hat{y}_i|^2\right)\right) \tag{15}$$

to minimise the Bayesian Information Criteria (BIC), the reduced χ^2 and outlier influence as per [38]. We report the values for ΔBIC [52], and the reduced χ^2 as χ^2/DOF where N is the number of data pairs, FP is the number of free parameters and DOF is the degree of freed-om, $DOF = N\text{-}FP$. We then apply nonlinear regression using the H-routine and distance values, D_L with Eqs. (3), (5), (7), (9), (10) and (13) as per [32] both with and without intercepts.

Using all 156 data pairs and (1,0) for the position of the local group, most regressions using the H-routine present results with $H_0 > 85$ km s^{-1} Mp^{-1} which we think unrealistic. The models present many shallow, local minima which is partly the result of the minuscule weights allowed the distant emissions with large associated errors. This makes model comparison difficult because unique, deep regression minima often cannot be found. We observe many values for D_L beyond the 99.99%

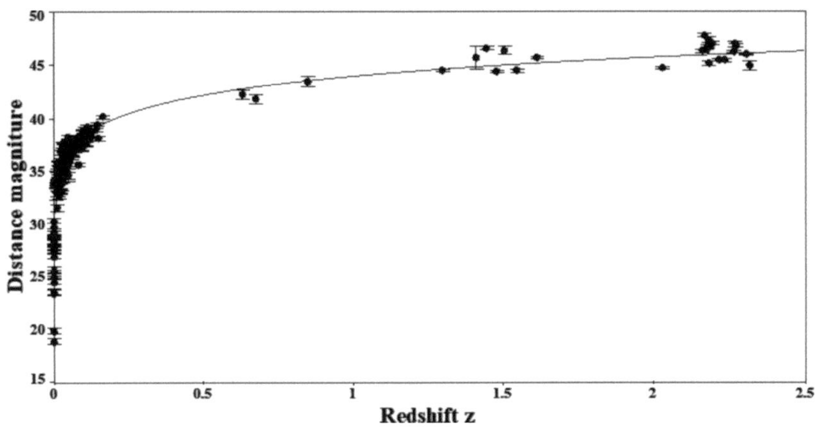

Figure 1.
Diagram of a pseudo-H-routine regression using all 156 HII and GEHR data. The line is the best fit of the ΛCDM model with H_0 *the prior at 70.9 km* s^{-1} *Mpc*$^{-1}$.

confidence interval with all models, and present data and results from regression with the ΛCDM model at H_0 = 70.9 in **Figure 1** as one example. There are 68 values of D_L which lie outside the 99.99% confidence limits ($\approx 4\sigma$). The standard deviations for many emission at $a < 0.86$ are a large fraction of the Universe diameter and many D_L values lie well above the 99.99% confidence limit when there should only be one or two. For these reasons we decided to parse the data using two different conservative methods.

3. Results

We can reproduce the results of [24] using the routine of correlating *mag* versus *z* with adjustments of nuisance parameters. The data handling routines are checked by first using all 156 HII/GEHR observations and the pseudo-H-regression routine against reported results with Eqs. (4), (6) and (8). We find parameters and goodness of fit values very similar to those of [24]. We present diagrams of our regression using the ΛCDM model using all 156 *mag* and redshift data pairs in **Figures 1** and 2. We chose to display results from the standard model even though this is not the best fit with these data but because this model is the most popular. Note the standard deviations of distant emissions appear similar to those of more nearby events which is unrealistic. It is common knowledge that distant objects are more difficult to measure accurately than those nearby and measuring luminosity through billions of years of intergalactic dust must introduce more noise than for nearby emissions.

This problem is highlighted in **Figure 2** where it is obvious the standard deviations are very similar for nearby emissions and those at $z > 1$, which have travelled more than 6 billion light years. Also note the ordinate intercept cannot be displayed as (0,0), which is the location and relative velocity of the local group. This most accurate data pair cannot be used with this diagram type. Diagrams such as **Figures 1** and 2 using SNe Ia data have been used for presentation by many continuing to this day [1, 9, 43, 44, 53].

When we attempt regression following the robust H-routine with Eqs. (3), (5), (7), (9), (10) and (13) using all 156 data pairs plus the local group position (1,0), we fail to find satisfactory solutions at $H_0 < 85$ km s^{-1} Mpc^{-1}; the high side of a realistic value. We attempt regression using many different data handling routines;

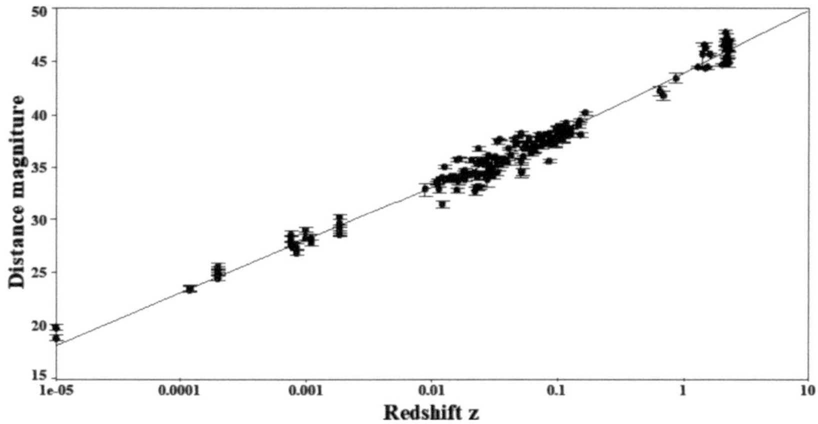

Figure 2.
Diagram of a pseudo-H-routine regression using all 156 HII and GEHR data. The centre line is the best fit of the ΛCDM model with H_0 fixed at 70.9 km s^{-1} Mpc^{-1}. Note the abscissa is log_{10} format to better present nearby emissions.

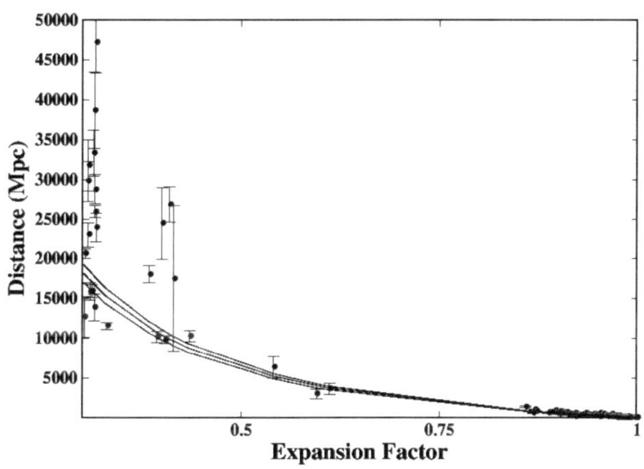

Figure 3.
The H-routine regression with all 157 HII and GEHR data plus 10 nearby galaxies. The centre line is the best fit for the Rh = ct model with H_0 of 65.9 km s^{-1} Mpc^{-1}. The outer lines are the ±99.99% confidence intervals. The high side of the SD of the emission at a of 0.316 is not presented.

anchoring the HII/GEHR data in several manners, testing the data as various large segments, data without those of $z > 0.18$ ($a < 0.85$) with fixed or floating GEHR values or with and without the GEHR data. We tried including the positions of 10 nearby galaxies to improve the results with but to no avail.

We present one example of our many attempts in **Figure 3**, where we include data from those 10 neighbouring galaxies with fixed distances [54]. This presents relatively small scatter about the best fit for data $a > 0.85$ but most values less than that are well above the best fit, that is, further distant than predicted and exhibiting large standard errors.

Examinations of both **Figures 3** and **4** reveal that nearby HII/GEHR sources display relatively small distance dispersion and errors, while ancient emissions are very scattered with very large estimated errors, as expected for difficult, distant observations. This presentation is very different from that displayed in **Figures 1** and **2**; the match of H_0 at 71 was made by 'massaging' the *mag* data, that is by adjusting other parameters in order to recalculate the distances and associated standard deviations as necessary. The scattering data at $a < 0.85$ [22], are the

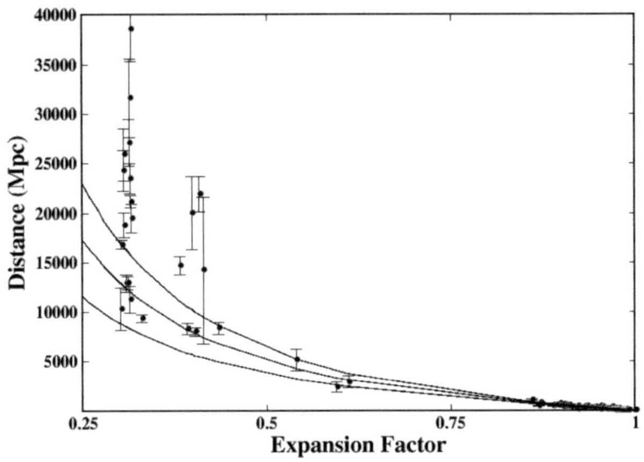

Figure 4.
Diagram of an H-routine regression using all 156 HII and GEHR data plus the local group (1,0). The centre line is the best fit of the ΛCDM model with H_0 fixed at 70.9 km s^{-1} Mpc^{-1}. The outer lines border the ±99.99% confidence interval. The high side of the standard deviation of the emission at a = 0.316 at 50,340 Mpc is not presented. The Big Bang is at expansion factor 0, the local group of galaxies including our Milky Way is at 1,0.

primary reasons for our inability to reach believable best fits for these models when we attempt regression without strong restrictions or 'priors'.

H_0 is the most important parameter for regression of FLRW-type models and is highly dependent on overall curvature of the regression, the slope if you will allow, and hence distant data quality. Because distant data are very noisy and suffer systematic error, these values are nearly ignored using weighed, computerised regression. The regression then ignores distant signals and becomes highly dependent on nearby SNe values. Unfortunately, this means the HII/GEHR data are currently of limited value for determining H_0 and other cosmological parameters. Investigators relying on the pseudo-H-routine as displayed by **Figures 1** and **2** may claim [22], we are now in the era of 'precision cosmology' but evidence in these figures says otherwise.

For model comparisons in **Table 1** we list results from robust H-routine regressions with H_0 of 71 as preferred by some working with HII/GEHR data [23]. Results are organised using the relative values of the calculated Bayesian information criteria (ΔBIC) [24] also with the reduced χ^2 values. The spread of both descriptors is much wider than calculated using the pseudo-H-routine [24], making

Model	ΔBIC	χ^2/(N-FP)	H_0[a]	Ω_m	Intercept (Mpc)
Rh = ct	0	23.95	71.0 ± 2.1	—	0.03
ωΛCDM	88.3	24.03	71.0 ± 8.3	1 ± >1000	0.03
CGR	91	21.9	70.7 ± 3.4	1 ± 0.75	0.07
ΛCDM	110	23.0	70.9 ± 2.9	1 ± 0.28	0
EdS	123	25.4	71.3 ± 2.4	1	0.03
ST	140	22.6	71.1 ± 10.8	1 ± 0.18	0

[a]*km s^{-1} Mpc^{-1}.*
ΔBIC are the relative values of the Bayesian information criteria and χ^2/(N-FP) is the reduced sum of χ^2 with FP the number of free parameters.

Table 1.
Analysis of D_L versus expansion factor with 157 observations including the local group (1.0) for regression minima close to 71 km s^{-1} Mpc^{-1}.

discrimination between models easier. Note all intercepts are negligible indicating little systematic error in nearby signals.

The results for the standard model are eclipsed by those of the Rh = ct model when all 156 HII/GEHR values are used with the H-routine, **Table 1**. These strongly support the findings of [24], that is, if judged by the lowest value of ΔBIC. If judged by the lowest value of χ^2 the CGR model best describes the HII/GEHR data. The two versions of the standard model, ΛCDM, ωΛCDM, perform poorly compared to the *Rh* = ct model. We are puzzled by the high values for Ω_m in the standard models which are much larger than found earlier using the H-routine with the SNe Ia data [46].

The results in **Table 2** are from data parsed using the 99.99% limits reducing the database by over 40%, though we consider this a conservative parse. The H-routine regressions for all models begins presuming an initial H_0 of 71. The values for Ω_m are higher than expected and both the *Rh* = ct and CGR models perform poorly describing these data. On the other hand, a version of the current standard model, ωΛCDM, performs best considering the ΔBIC values but not significantly better than the ST model if one considers the reduces χ^2 values.

The results in **Table 3** are from data parsed using the Studentised limit discarding values of $ri/\sigma i > 1.5$, which reduces the database a bit further. The ωΛCDM, model only slightly outperforms the ΛCDM and ST models with the *Rh* = ct model performing poorly. The values for Ω_m are all again much higher than expected. Using these parsed data and Eqs. (3), (5), (7), (9), (10) and (13) we can easily discriminate between these more popular models based on the values

Model	ΔBIC	χ^2/(N-FP)	H_0[a]	Ω_m	Intercept (Mpc)
ωΛCDM	0	6.04	65.7 ± 4.4	1 ± >1000	−0.07
ST	25.7	5.90	76.0 ± 9	1 ± 0.29	0
ΛCDM	27.8	5.98	76.4 ± 2.4	1 ± 0.18	0
CGR	28.3	8.11	69.2 ± 1.9	1 ± >1000	0
Rh = ct	43.4	7.22	73.8 ± 1.6	—	0.03
EdS	54.4	6.68	70.7 ± 1.7	1	0.03

[a]*km* s^{-1} *Mpc*$^{-1}$.

Table 2.
Results from D_L versus expansion factor with parsed observations including the local group within 99.99% confidence, 89 data pairs.

Model	ΔBIC	χ^2/(N-FP)	H_0[a]	Ω_m	Intercept (Mpc)
ωΛCDM	0	2.78	66.0± >1000	1 ± >1000	0.02
ΛCDM	1.8	2.49	69.0 ± 1.6	1 ± 0.15	0.02
ST	2.9	2.49	69.5 ± 1.4	1 ± 0.16	0.02
EdS	21.3	3.10	66.1 ± 1.3	1	0.02
Rh = ct	36.4	4.58	66.0 ± 1.	—	0.02
CGR	41	5.13	62.3 ± 2.1	1 ± 0.53	0

[a]*km* s^{-1} *Mpc*$^{-1}$.

Table 3.
Results from D_L versus expansion factor with parsed observations including the local group using a studentised limit of 1.5, 74 data pairs.

calculated for ΔBIC. As expected, the EdS model is too simple and never a good description of the HII/GEHR.

4. Conclusions and discussion

There is a current controversy around the best general description of our Universe. The popular ΛCDM and ωΛCDM models rely heavily on SNe Ia, Cepheid variable and CMB data for validity as per Riess et al. [9]. Another, the *Rh* = ct (the eternal, coasting, non-empty) model functions slightly better than the former two models when tested by proponents with the same SNeIa data as reported by Wei et al. [55] and with HII/GEHR data [24]. We can reproduce the results of this latter group using their selected data by following the pseudo-H-routine. We acknowledge a serious effort has been made by them to analyse these data using their best techniques, the pseudo-H-routine. Unfortunately their analytical method is flawed, as we have pointed out in our Introduction, leading to questionable results and conclusions by many groups.

We first employ all 156 data pairs organised by Wei et al. [24] with the local group as the origin (1,0) using the H-routine; **Table 1**. Examining **Figures 3** and **4**, the distant data, $a < 0.85$, are too scattered with large errors to trust our results so we are forced to use a prior for H_0 as 71 km s^{-1} Mpc^{-1} We find the *Rh* = ct and CGR models describe the HII/GEHR unparsed data better than the current standard models, ΛCDM and ωΛCDM. To properly evaluate using the H-routine we are forced to parse the data; this enables the regression procedure to find proper minima. Analyses with the parsed data supports a lower value for H_0 than that of Riess [1], but does not prefer the *Rh* = ct model rather the current standard models; **Tables 2** and **3**.

One reason for the discrepancy between parsed and unparsed ensembles is the large dispersion of HII data with large errors for HII distances at $a < 0.85$ as shown in **Figures 3** and **4**. These large errors are automatically hidden and their influence on regression is drastically increased when the pseudo-H-routine is used, **Figures 1** and **2**. We think the results and conclusions of the Riess [9] and Wei groups [24] are tainted by this type of analysis. If the analyses by these groups be useful and if the pseudo-H-routine be a valid method, our results using the H-routine and the pseudo-H-routine should be similar, but are not [16]. We wonder if the *Rh* = ct is really a useful model, since solutions do not present values for cosmological parameters other than H_0. The *Rh* = ct model has other problems as well [28, 36].

The results in **Tables 2** and **3** should not be taken as endorsing the standard model. First, we think the value of the HII/GEHR data, especially events older than $a \approx 0.85$ is suspect. Second, we have previously shown the complete Einstein field equation, including Λ, when modelled by the FLRW conditions, displays mathematical inconsistencies incompatible with reality [56]. Third, we have recently shown that even Λ tuned to Universe expansion, or tuned to the Hubble-Lemaitre constant, or dependent on decreasing matter density with increasing time, cannot rectify the fundamental problems with that concept [57]. There we have shown by tracing the value of Ω_Λ back towards recombination demands ridiculous values for either Ω_m or Ω_k or both. Fourth, the results presented here from analyses of the parsed data with the ST model are as good as the standard model. Finally, an attempt to measure dark energy as a new force failed a sensitive laboratory test [51].

Our picture of the Universe is complicated; when the ΛCDM model is assayed with CMB data, H_0 is significantly lower (66.9 km s^{-1} Mpc^{-1}) than calculated using SNe Ia data (74.2), both with small claimed errors; [1, 3, 9] but the opposite is expected in a universe suffering gravity. Both the CMB value for H_0 and the evaluation procedure using that data have been recently, vigorously contested by Riess [7]. In addition to those published arguments, we note that analysis of the CMB data relies on 6 parameters with many required priors, using signal averaged data produced at only a single, distant moment. These are discussion points which are rarely mentioned but which we feel severely weakens the value of the CMB results [11]. On the other hand, we have previously pointed out the SNe Ia data are very noisy, much like the HII/GEHR data shown here [46, 58]. When these data are evaluated with a questionable technique using a 4 or 5 parameter regression in reality, it is not surprising the results from using SNe Ia or HII/GEHR data as standard candles do not always match those of other groups; results from LIGO/Virgo [5, 6].

But why do not astronomers and physicists realise and correct this mistake? The analysis of SNe Ia and HII/GEHR is difficult and time-consuming, thousands of readers prefer to simply trust the results and conclusions of articles written by well-known groups rather than take time and brain-power re-investigating the analyses. But why do astronomers persist in using a system, *mag* versus z, which does not yield meaningful results? First, because this is the manner astrophysics was taught and is still followed. Decades ago scientists were forced to plot data in semilog or log-log formats to calculate a value for a rate constant, for instance following a chemical reaction, bygone days when any value was better than none. This practice has been superseded by the use of high-speed computers which can better model relevant data and better calculate rate constants—in our case the Hubble-Lemaitre constant. Indeed it has been decades since semilog or log-log plots have been tolerated in biophysics [59]. Astronomy students are not being taught the better techniques of data analysis but the older methods of semilog or log-log plot. Second, data are often organised in tables of redshift and distance magnitude (μ). Both measures are used for historical reasons, magnitude being related to luminosity as perceived by the human eye being approximately the log(luminosity). Some astronomers actually worry more about relatively minor redshift errors than investigate the larger distance errors [37]. Third, results and parameters from this type of analysis (pseudo-H-routine) are interesting and immediately useful in today's astrophysics. The concept of an accelerating Universe expansion due to the release of something like *dark energy* from the spacetime between galactic groups gives the thrill of discovery to astrophysics. This new concept is justification for the billions of $ spent on large telescopes and satellites, otherwise the money only bought pretty pictures. Fourth, the hope for 'new' physics to young students. This new concept, dark energy, is now fertile ground for theoretical physicists with hundreds of articles published the past 20 years. As a side-note, dozens of articles proposing versions of the MOND model (Modified Newtonian Dynamics) are now defunct, having been dis-proven by the simultaneous observations of light and gravity waves from a binary neutron star collision [60–62]. Fifth, the concept of dark energy has hatched projects employing dozens of astronomers and several large telescopes [63, 64]. Many astronomers are now dedicating time to DES, the Dark Energy Survey, presuming the ΛCDM model correctly describes our Universe [65]. Sixth, many cosmologists and astrophysicist should change their analytical procedures but will not [16]. Finally, the discovery of accelerating Universe expansion has been ennobled with the highest prize for physics [66].

The tension over the correct value of H_0 might be resolved if another set of standard candles, independent of SNe Ia and gamma-ray burst emissions and stretching beyond $z \approx 2$ could be used for independent model testing with the correct analytic technique, for example, a better quality HII/GEHR data set. Another reason why independent data are needed is because those working with SNe Ia data present the regression for the ΛCDM model as requiring only 2 or 3 parameters; this is really a 4 or 5 parameter regression. (Because the distance data and H_0 must be adjusted between attempted regressions and between models another 2 parameters are introduced making regression a 4 or 5 parameter pursuit. One might term this *data massage*.) In reality, large ensembles of noisy SNe Ia data are only slightly better tools for model discrimination than the ambiguous, 6 parameter fits with CMB data and both SNe Ia and CMB analyses suffer overfitting [11]. Independent data, perhaps from red giants [17], are also important to test the many varieties of dark energy and exotic models now hypothesised to explain the SNe Ia data. These are good reasons why emissions from HII galaxies, GEHR and red giant stars should be seriously considered and encouraged. Extra effort should now be made collecting and analysing many more and better HII/GEHR and red giant signals, the sooner the better, for these data offers a truly independent means of estimating important cosmological parameters. We encourage those collecting and analysing SNe and HII/GEHR data to give more thought to better data analysis and to consider more than just their favourite model.

Conflict of interest

The authors declare that there exists no conflict of interest.

Author details

Michael L. Smith[1*] and Ahmet M. Öztaş[2]

1 Umeå, Sweden

2 Department of Physics Engineering, Hacettepe University, Ankara, Turkey

*Address all correspondence to: mlsmith55@gmail.com

References

[1] Riess AG et al. A 2.4% determination of the local value of the Hubble constant. The Astrophysical Journal. 2016;**826**:56. DOI: 10.3847/0004-637X/826/1/56

[2] Abbott TMC et al. First cosmology results using type Ia supernovae from the dark energy survey: Constraints on cosmological parameters. The Astrophysical Journal. 2019;**872**:L30. DOI: 10.3847/2041-8213/ab04fa

[3] Ade PAR et al. Planck 2015 results—XIII. Cosmological parameters. Astronomy and Astrophysics. 2016;**594**:A13. DOI: 10.1051/0004-6361/201525830

[4] Aghanim N et al. Planck 2018 Results. VI. Cosmological Parameters. 2018. Available from: arXiv:1807.06209 [Accessed: 20 September 2019]

[5] LIGO Scientific Collaboration and Virgo Collaboration et al. A gravitational-wave standard siren measurement Hubble constant. Nature. 2017;**551**:85-88. DOI: 10.1038/nature24471

[6] Soares-Santos M et al. First measurement of the Hubble constant from a dark standard siren using the dark energy survey galaxies and the LIGO/Virgo binary-black-hole merger GW170814. Astrophysical Journal Letters. 2019;**876**:L7. DOI: 10.3847/2041-8213/ab14f1

[7] Riess AG et al. Seven Problems with the Claims Related to the Hubble Tension in arXiv:1810.02595. 2018. Available from: arXiv:1810.03526

[8] Shanks T, Hogarth L, Metcalfe N. *Gaia* cepheid parallaxes and 'Local Hole' relieve H_0 tension. Monthly Notices of the Royal Astronomical Society. 2018;**484**:L64-L68. DOI: 10.1093/mnrasl/sly239

[9] Riess AG et al. Large magellanic cloud cepheid standards provide a 1% foundation for the determination of the Hubble constant and stronger evidence for physics beyond ΛCDM. The Astrophysical Journal. 2019;**876**:85. DOI: 10.3847/1538-4357/ab1422

[10] skydivephil. Cosmology's Latest Puzzle: The Hubble Tension Featuring Adam Reiss. 2018. Available from: https://www.youtube.com/watch?v=uoAkFq-KIrk

[11] Coolen ACC et al. Replica analysis of overfitting in regression models for time-to-event data. Journal of Physics A: Mathematical and Theoretical. 2017;**50**:375001. DOI: 10.1088/1751-8121/aa812f

[12] Kessler R, Scolnic D. Correcting type Ia supernova distances for selection biases and contamination in photometrically identified samples. The Astrophysical Journal. 2017;**836**:56. DOI: 10.3847/1538-4357/836/1/56

[13] Riess AG et al. New parallaxes of galactic cepheids from spatially scanning the Hubble space telescope: Implications for the Hubble constant. The Astrophysical Journal. 2018;**855**:136. DOI: 10.3847/1538-4357/aaadb7

[14] Huang CD et al. A near-infrared period-luminosity relation for Miras in NGC 4258, an anchor for a new distance ladder. The Astrophysical Journal. 2018;**857**:67. DOI: 10.3847/1538-4357/aab6b3

[15] Vishwakarma RG, Narlikar JV. Is it no longer necessary to test cosmologies with type Ia supernovae? Universe. 2018;**4**:73. Available from: https://www.mdpi.com/2218-1997/4/6/73

[16] Smith ML, Oztas AM. Log-transformation Problems. 2017. Available from: https://www.youtube.com/watch?v=Y1nEQmg2yJA&t=120s

[17] Freedman WL et al. The Carnegie-Chicago Hubble program. VIII. An independent determination of the Hubble constant based on the tip of the red giant branch. The Astrophysical Journal. 2019;**882**:34. DOI: 10.3847/1538-4357/ab2f73

[18] Beaton RL et al. The Carnegie-Chicago Hubble Program. VII. The Distance to M101 via the Optical Tip of the Red Giant Branch Method. Available from: arXiv:1908.06120

[19] Chavez R et al. Determining the H_0 using giant extragalactic HII regions and HII galaxies. Monthly Notices of the Royal Astronomical Society. 2012;**425**: L56-L60. DOI: 10.1111/j.1745-3933.2012.01299.x

[20] Chavez R et al. The Lσ relation for massive bursts of star formation. Monthly Notices of the Royal Astronomical Society. 2014;**442**: 3565-3597. DOI: 10.1093/mnras/stu987

[21] Chavez R et al. Constraining the dark energy equation of state with HII galaxies. Monthly Notices of the Royal Astronomical Society. 2016;**462**: 2431-2439. DOI: 10.1093/mnras/stw1813

[22] Terlevich R et al. On the road to precision cosmology with high-redshift HII galaxies. Monthly Notices of the Royal Astronomical Society. 2015;**451**: 3001-3010. DOI: 10.1093/mnras/stv1128

[23] Fernandez Arenas DF et al. An independent determination of the local Hubble constant. Monthly Notices of the Royal Astronomical Society. 2017; **474**:1250-1276. DOI: 10.1093/mnras/stx2710

[24] Wei JJ, Wu XF, Melia F. The HII galaxy Hubble diagram strongly favours Rh = ct over ΛCDM. Monthly Notices of the Royal Astronomical Society. 2016; **463**:1144-1152. DOI: 10.1093/mnras/stw2057

[25] Leaf K, Melia F. A two-point diagnostic for the HII galaxy Hubble diagram. Monthly Notices of the Royal Astronomical Society. 2017;**474**: 4507-4513. DOI: 10.1093/mnras/stx3109

[26] Yennapureddy MK, Melia F. Reconstruction of the HII galaxy Hubble diagram using gaussian processes. Journal of Cosmology and Astroparticle Physics. 2017;**11**:29. DOI: 10.1088/1475-7516/2017/11/029

[27] John MV, Joseph KB. Generalized Chen-Wu type cosmological model. Physical Review D. 2000;**61**:087304. DOI: 10.1103/PhysRevD.61.087304

[28] John MV. Rh = ct and the eternal coasting cosmological model. Monthly Notices of the Royal Astronomical Society. 2019;**484**:L35-L37. DOI: 10.1093/mnrasl/sly243

[29] Melia F, Yennapureddy MK. Model selection using cosmic chronometers with gaussian processes. Journal of Cosmology and Astroparticle Physics. 2018;**2**:34. DOI: 10.1088/1475-7516/2018/02/034

[30] Sandage A et al. The Hubble constant: A summary of the Hubble space telescope program for the luminosity calibration of type Ia supernovae by means of cepheids. The Astrophysical Journal. 2006;**653**:843. DOI: 10.1086/508853

[31] Zheng J, Melia F, Zhang TJ. A Model-independent Measurement of the Spatial Curvature using Cosmic Chronometers and the HII Hubble Diagram. 2019. Available from: arXiv:1901.05705

[32] Hubble E. A relation between distance and radial velocity among extra-galactic nebulae. Proceedings of the National Academy of Sciences of the

United States of America. 1929;**15**: 168-173. DOI: 10.1073/pnas.15.3.168

[33] Feng C et al. Log-transformation and its implications for data analysis. Shanghai Archives of Psychiatry. 2014; **26**:105-109. DOI: 10.3969/j.issn. 1002-0829.2014.02.009

[34] Packard GC, Birchard TJ, Boardman TJ. Fitting statistical models in bivariate allometry. Biological Reviews. 2011;**86**: 549-563. DOI: 10.1111/j.1469-185X. 2010.00160.x

[35] Packard GC. Relative growth by the elongated jaws of gars: A perspective on polyphasic loglinear allometry. Journal of Experimental Zoology. Part B, Molecular and Developmental Evolution. 2016;**326B**:168-175. DOI: 10.1002/jez.b.22673

[36] Oztas AM, Smith ML. Spacetime curvature and the cosmic horizon: Derivations using the Newtonian world and the Friedmann-Robertson-Walker metric. Monthly Notices of the Royal Astronomical Society. 2015; **449**:1270-1274. DOI: 10.1093/mnras/ stv346

[37] Davis TM et al. Can redshift errors bias measurements of the Hubble constant? Monthly Notices of the Royal Astronomical Society. 2019; **490**:2948-2957. DOI: 10.1093/mnras/ stz2652

[38] Riazoshams H, Midi H, Gebrenegus G. Robust Nonlinear Regression with Applications Using R. UK: John Wiley; 2019. p. 264

[39] Carroll SM, Press WH, Turner EL. The cosmological constant. Annual Review of Astronomy and Astrophysics. 1992;**30**:499-542. DOI: 10.1146/annurev. aa.30.090192.002435

[40] Carroll SM. The cosmological constant. Living Reviews in Relativity. 2001;**4**:1. DOI: 10.12942/lrr-2001-1

[41] Piattella O. Lecture Notes in Cosmology. UK: Springer Nature; 2018. DOI: 10.1007/978-3-319-95570-4

[42] Melia F, Shevchuk ASH. The Rh = ct universe. Monthly Notices of the Royal Astronomical Society. 2011;**419**: 2579-2586. DOI: 10.1111/j.1365-2966. 2011.19906.x

[43] Riess AG et al. Observational evidence from supernovae for an accelerating universe and a cosmological constant. Astronomy Journal. 1998;**116**: 1009-1038. DOI: 10.1086/300499

[44] Perlmutter S et al. Measurements of Ω and Λ from 42 high-redshift supernovae. The Astrophysical Journal. 1999;**517**:565-586. DOI: 10.1086/307221

[45] O'Raifeartaigh C, Mitton S. Interrogating the legend of Einstein's "Biggest Blunder". Physics in Perspective. 2018;**20**:318-341. DOI: 10.1007/s00016-018-0228-9

[46] Oztas AM, Smith ML, Paul J. Spacetime curvature is important for cosmology constrained with supernova emissions. International Journal of Theoretical Physics. 2008;**47**:2464-2478. DOI: 10.1007/s10773-008-9680-7

[47] Carmeli M. Relativity: Modern Large Scale Spacetime Structure of the Cosmos. Singapore: World Scientific; 2008. DOI: 10.1142/6820 ISBN 978-981-281-375-6

[48] Hartnett JG. Spiral galaxy rotation curves determined from Carmelian general relativity. International Journal of Theoretical Physics. 2006;**45**:2118-2136. DOI: 10.1007/s10773-006-9178-0

[49] Schumann M. Direct detection of WIMP dark matter: Concepts and status. Journal of Physics G: Nuclear and Particle Physics. 2019;**46**:103003. DOI: 10.1088/1361-6471/ab2ea5

[50] Roberts G Jr. New Results from World's Most Sensitive Dark Matter

Detector. Berkeley Lab News Center; 2015. Available from: https://newscenter.lbl.gov/2015/12/14/new-lux-results/-results/

[51] Sabulsky DO et al. Experiment to detect dark energy forces using atom interferometry. Physical Review Letters. 2019;**123**:061102. DOI: 10.1103/physrevlett.123.061102

[52] Schwarz G. Estimating the dimension of a model. The Annals of Statistics. 1978;**6**:461-464. Available from: https://projecteuclid.org/euclid.aos/1176344136

[53] Kirshner RK. Hubble's diagram and cosmic expansion. Proceedings of the National Academy of Sciences of the United States of America. 2003;**101**: 8-13. DOI: 10.1073/pnas.2536799100

[54] Sandage A, Bedke J. Atlas of Galaxies, NASA SP-496, Library of Congress Cataloging-in-Publication Data; 1988

[55] Wei JJ et al. A comparative analysis of the supernova legacy survey sample with ΛCDM & Rh = ct universe. Astronomy Journal. 2015;**149**:102. DOI: 10.1088/0004-6256/149/3/102/meta

[56] Oztas AM, Smith ML. Elliptical solutions to the standard cosmology model with realistic matter density. International Journal of Theoretical Physics. 2006;**45**:896-907. DOI: 10.1007/s10773-006-9082-7

[57] Oztas AM, Dil E, Smith ML. The varying cosmological constant: A new approximation to the Friedmann equations and universe model. Monthly Notices of the Royal Astronomical Society. 2018;**476**:451-458. DOI: 10.1093/mnras/sty221

[58] Oztas AM, Smith ML. The cosmological constant constrained with Union2.1 supernovae type data. International Journal of Theoretical Physics. 2014;**53**:2636-2661. DOI: 10.1007/s10773-014-2061-5

[59] Ackers GK. President, American Biophysical Society. Personal communication; 1980

[60] Boran S et al. GW170817 falsifies dark matter emulators. Physical Review D. 2017;**97**:4. DOI: 10.1103/physrevd.97.041501

[61] Ezquiaga JM, Zumalacarregui M. Dark energy after GW170817: Dead ends and the road ahead. Physical Review Letters. 2017;**119**:251304. DOI: 10.1103/physrevlett.119.251304

[62] Lee C. Colliding Neutron Stars Apply Kiss of Death to Theories of Gravity. 2017. Available from: https://arstechnica.com/science/2017/10/colliding-neutron-stars-decapitate-zombie-theory-of-gravity/

[63] Abbott TMC et al. Dark energy survey year 1 results: A precise H_0 estimate from DES Y1, BAO, and D/H data. Monthly Notices of the Royal Astronomical Society. 2018;**480**: 3879-3888. DOI: 10.1093/mnras/sty1939

[64] Abbott TMC et al. Cosmological constraints from multiple probes in the dark energy survey. Physical Review Letters. 2019;**122**:17. DOI: 10.1103/physrevlett.122.171301

[65] Abbott TMC et al. Dark energy survey year 1 results: Cosmological constraints from galaxy clustering weak lensing. Physical Review D. 2018;**98**: 043526. DOI: 10.1103/physrevd.98.043526

[66] Riess AG. Nobel lecture: My path to the accelerating universe. Reviews of Modern Physics. 2012;**84**:1165-1175. DOI: 10.1103/revmodphys.84.1165

Chapter 4

Nature of Dark Energy

Jan Olof Stenflo

Abstract

When supernova observations in the end of the 1990s showed the cosmic expansion to be accelerating, it became necessary to reintroduce the cosmological constant Λ as a fitting parameter. Although its physical origin has remained a mystery, it has generally been interpreted as some kind of energy field referred to as "dark energy." This interpretation however implies a cosmic coincidence problem because we happen to live at a time when dark energy becomes the dominant driver of the expansion. Here we present an alternative explanation: The Λ term is induced by a global boundary constraint that ties its value to the conformal age of the universe. The cosmic coincidence problem then goes away. We illustrate how the cosmological evolution that is implied by this constraint differs from standard cosmology. Without the use of any free parameters, the theory predicts a present value of Λ that is within 2σ from the value derived from CMB observations with the Planck satellite. The universe is found to be mildly inflationary throughout the entire radiation-dominated era. This obviates the need to postulate a hypothetical, violent grand unification theory (GUT) era inflation to explain the observed large-scale homogeneity and isotropy of the universe.

Keywords: dark energy, cosmology, theory, inflation, gravitation, early universe

1. Introduction

The term "dark energy" refers to the cosmological constant Λ when interpreted as some kind of mysterious energy field that pervades space and exerts a negative pressure, which is the source of the observed accelerated expansion of the universe. Einstein [1] introduced the cosmological constant in 1917 to allow for a static universe but considered it a blunder after the cosmic expansion was discovered. It was only after the discovery of the accelerated expansion in the end of the 1990s through the use of supernovae type Ia as standard candles [2, 3] that it became necessary to reintroduce Λ as a fitting parameter to allow the observations to be modeled. Its physical nature has however been enigmatic and elusive. In particular the observed magnitude of Λ appears to make us "privileged observers," because we happen to live at a time when dark energy starts to dominate over the energy densities of matter and radiation, thereby causing the onset of an inflationary phase of the universe that will continue forever. This is often referred to as the "cosmic coincidence problem."

Dark energy is widely regarded as one of the biggest problems in contemporary physics (for a review, cf. [4]). All conceivable ways to modify gravity have been tried. Different approaches to model the observational data have been explored, e.g. [5]. Elaborate laboratory experiments have been performed in the search for new scalar fields that would modify gravity [6]. On top of this, evidence against the earlier interpretation of the supernova observations in terms of dark energy has been discovered [7].

Recently [8] it was shown that there is an alternative way to explain the need for a cosmological constant, namely, as the result of a global cosmic boundary constraint instead of through the introduction of some new physical field. This approach leads to a new cosmological framework that brings a resolution to several outstanding enigmas, including the cosmic coincidence problem. Without the use of any free parameters, Λ is predicted to have a present value that is within 2σ from the value that has been determined from CMB data with the Planck satellite [9]. The evolution of the scale factor that is derived with the new theoretical framework shows that the universe has been in a mildly accelerating, inflationary phase throughout all of the radiation-dominated era since the beginning of the Big Bang. This automatically explains the observed large-scale homogeneity and isotropy of the universe without any need to postulate a hypothetical violent inflationary phase in the grand unification theory (GUT) era of the early universe.

In Section 2 we review the arguments that have been presented in [8] for the origin of the global constraint that governs the value of Λ. These arguments depend on the participatory role of observers in the universe for the needed definition of cosmic time, with the split between past and future and the distinction between dynamic time and nonlocal (look-back) time. This will be clarified in Section 3. The mathematical equations of the new cosmological framework are derived and solved in Section 4, where we also illustrate how the cosmic evolution differs from that of standard cosmology. In Section 5 we show how inflation emerges as a natural part of cosmic history throughout the radiation-dominated era, thereby eliminating the causality problem without the assumption of any new fields. The conclusions are summarized in Section 6.

2. Resonant origin of the Λ term

In standard cosmological models, the universe is assumed to be homogeneous and isotropic on the largest scales, because this is what observations tell us. The cosmological evolution can then conveniently be described in terms of a scale factor $a(t)$ that only depends on time t. If we further assume zero spatial curvature, the metric can be expressed as

$$\mathrm{d}s^2 = -c^2\mathrm{d}t^2 + a(t)^2\left(\mathrm{d}r^2 + r^2\,\mathrm{d}\Omega\right), \tag{1}$$

where r is the comoving distance and $\mathrm{d}\Omega$ is the surface element on the unit sphere. While observations show that there is no significant spatial curvature at the present epoch, there is also a theoretical justification for the validity of the flatness assumption, which emerges within the framework of the alternative cosmology of the present work. This will be clarified in Section 4.2.

Besides "proper time" t, we will need to make use of two other time concepts: "conformal time" η and "Euclidian conformal time" τ. The relation between them is defined by

$$\mathrm{d}\tau \equiv ic\,\mathrm{d}\eta \equiv ic\,\mathrm{d}t/a. \tag{2}$$

In terms of the temporal coordinates η and τ, the metric becomes

$$\begin{aligned} \mathrm{d}s^2 &= a(\eta)^2\left(-c^2\mathrm{d}\eta^2 + \mathrm{d}r^2 + r^2\,\mathrm{d}\Omega\right), \\ \mathrm{d}s^2 &= a(\tau)^2\left(\mathrm{d}\tau^2 + \mathrm{d}r^2 + r^2\,\mathrm{d}\Omega\right). \end{aligned} \tag{3}$$

The conformal metric of the first of these two equations shows that the metric coefficients are proportional to $\eta_{\mu\nu}$, the Minkowski metric: $g_{\mu\nu} = a^2\eta_{\mu\nu}$. "Conformal" means that all angles and shapes of trigonometric functions are preserved in spite of the nonlinear temporal dependence of the scale factor a. Fourier decompositions are only meaningful within the conformal framework.

The word "Euclidian" as the term for the second metric in Eq. (3) does not refer to the flatness assumption but to the signature of the metric: $(++++)$ instead of the $(-+++)$ signature when using time t or η. Since the τ coordinate then formally behaves like a spatial coordinate, we have incorporated the speed of light c in the definition of τ in Eq. (2), to let τ have the dimension of space.

The transformation to Euclidian spacetime leads to remarkable advantages and insights, which have found important applications in various areas in the form of Euclidian field theory, e.g., in solid-state physics [10]. The Hamiltonian in ordinary spacetime becomes the Lagrangian in Euclidian spacetime. Quantum field theory QFT in Euclidian spacetime has the structure of statistical mechanics in ordinary spacetime. The oscillating phase factors in QFT become the Boltzmann factors, while the path integral becomes the partition function. Euclidian spacetime has long been known to provide a direct and elegant route to the derivation of the Hawking temperature of black holes, cf. [8, 11].

In the following we will show how the oscillating phase factors of the Euclidian metric field contain a resonance that fixes the value of the cosmological constant Λ. Our starting point is the Einstein equation with cosmological constant, written in the form

$$R_{\mu\nu} - \Lambda g_{\mu\nu} = \frac{8\pi G}{c^4}\left(T_{\mu\nu} - \frac{1}{2}g_{\mu\nu}T\right). \tag{4}$$

This is the appropriate form to be used with the weak-field approximation, because the right-hand side of Eq. (4) represents the source term for gravitational waves when making a Fourier expansion, while the left-hand side describes the evolution of the vacuum fields, cf. [12]. We have here adopted the standard sign convention with $(-+++)$ for the spacetime signature and a plus sign in front of the right-hand side.

In the weak-field approximation and the harmonic gauge, $R_{\mu\nu} \approx -\frac{1}{2}\partial^2 g_{\mu\nu}$. The d'Alembertian operator $\partial^2 \equiv \Box^2 \equiv -(1/c^2)\partial^2/\partial t^2 + \nabla^2$. For the metrics of Eq. (3) the nabla operator in the d'Alembertian vanishes, because the spatial gradients can be disregarded on cosmological scales. The vacuum fields that represent the left-hand side of Eq. (4) then have the following weak-field representations in terms of the coordinates η and τ:

$$R_{\mu\nu} - \Lambda g_{\mu\nu} \approx \frac{1}{2c^2}\frac{\partial^2 g_{\mu\nu}}{\partial \eta^2} - \Lambda g_{\mu\nu} = -\frac{1}{2}\left(\frac{\partial^2 g_{\mu\nu}}{\partial \tau^2} + \frac{\omega_\Lambda^2}{c^2}\, g_{\mu\nu}\right). \tag{5}$$

While the vacuum fields without physical sources (the $T_{\mu\nu}$ fields) describe a de Sitter exponential evolution of the scale factor a when ordinary conformal time η is used as the temporal coordinate, they describe an anti-de Sitter-like universe when the τ coordinate is instead used. In this description the exponential evolution gets replaced by oscillating phase factors. To make it explicit that the τ representation leads to oscillating solutions, we have expressed it in terms of the oscillation frequency ω_Λ to give it the form of the equation for a harmonic oscillator. We have divided the frequency with c in Eq. (5) to make it a wave number, because τ was defined in terms of spatial units for reasons of symmetry with respect to the other three Euclidian coordinates. Nevertheless it is more appropriate to refer to the resonance in terms of a temporal frequency ω_Λ rather than a wave number, because it turns out to be related to the bounded nature of the observable timeline.

With the period of the oscillation given by $\eta_\Lambda = 2\pi/\omega_\Lambda$, the relation between the cosmological constant Λ, the oscillation frequency ω_Λ, and the period η_Λ, immediately follows from Eq. (5):

$$\Lambda = \frac{1}{2}\frac{\omega_\Lambda^2}{c^2} = 2\left(\frac{\pi}{c\eta_\Lambda}\right)^2. \tag{6}$$

In standard cosmology Λ in Eq. (4) is generally moved to the right-hand side, where it can be interpreted as a mass-energy density ρ_Λ. It is convenient to describe it in terms of the dimensionless parameter Ω_Λ, which is the fraction of the critical density ρ_c that is contributed by the Λ term:

$$\rho_\Lambda \equiv \Omega_\Lambda \rho_c = \frac{c^2\Lambda}{8\pi G}. \tag{7}$$

ρ_c represents the mean mass density that defines the boundary between open and closed model universes according to the Friedmann equations.

$$\rho_c = \frac{3}{8\pi G t_H^2}, \tag{8}$$

where $t_H = 1/H$ is the Hubble time and H the Hubble constant.
From Eqs. (6)–(8) we obtain

$$\frac{\eta_\Lambda}{t_H} = \frac{2\pi}{\sqrt{6\Omega_\Lambda}}. \tag{9}$$

Inserting the value of $\Omega_\Lambda = 0.685$ determined from observations with the Planck satellite [9], we get $\eta_\Lambda/t_H \approx 3.10$. As will be explicitly confirmed by the numerical solutions in Section 4, this implies a value of η_Λ that is nearly identical to the current conformal age η_u of the universe. The distance $r_u = c\eta_u$ is the radius of the particle horizon, the maximum distance to which an observer is causally connected. η_u is the time that it would take for a photon to travel this distance if the universe would stop expanding. As the spatial points from which light is emitted continually recede from us due to the cosmic expansion, η_u is substantially larger than the "proper age" t_u of the universe.

2.1 Link between Λ and the age of the universe

In standard cosmology Λ is a constant that should have nothing to do with the current age of the universe. This is contradicted by our finding that $\eta_\Lambda \approx \eta_u$. Any other value would be in conflict with the observed magnitude of the cosmic acceleration. If Λ were independent of the age of the universe, then η_Λ would have been many orders of magnitude larger than η_u in the past and will be many orders of magnitude smaller in the future. It would then be an extraordinarily improbable coincidence if they happen to be the same in the present epoch. This gives us strong reasons to suspect that the value of Λ is indeed physically tied to the age η_u of the universe.

The existence of such a physical link means that we need to single out, among all the solutions of the oscillator equation in Eq. (5), the Fourier component with a wavelength that corresponds to the conformal age of the universe. This only makes sense if time is bounded between the Big Bang and the Now, which seems to contradict the Einsteinian view that all future times somehow "already preexist" and that the experienced split between past, present, and future is just a stubborn illusion. Here we will argue (for details, see Section 3) that the Einsteinian view only makes sense in a universe devoid of observers and that this is not the universe that we inhabit. Like in quantum physics the observer plays a fundamental role in defining the nature of reality. The split between past, present, and future is not some illusion that we need to come to terms with, but is deeply physical. As soon as we introduce an observer (which can be a test particle, without brains or consciousness!) in Einstein's universe, the split occurs. In any observable universe the future does not exist, even in principle. The only accessible region is between the Big Bang and the Now, and this region is bounded. The theory has to be applied to the observable universe, not to some idealized universe without observers. This is not just some alternative philosophical viewpoint but has profound physical consequences. It leads to a very different cosmological framework, as will be made clear in the following sections.

The existence of a metric resonance with respect to Euclidian time τ implies that

$$g_{\mu\mu} \sim a(\tau)^2 \sim e^{\pm i\,\omega_u\,\tau/c}. \tag{10}$$

Note that the Euclidian metric and scale factor have here been treated like a quantum field by allowing them to have an analytical continuation into the complex plane. When we however convert back to ordinary conformal time η by replacing τ with $ic\eta$, the oscillating phase factor transforms into an exponentially evolving factor and thereby becomes real-valued. Both exponentially decaying and increasing solutions are possible because of the $\pm$ in Eq. (10). With the initial boundary condition that the scale factor was small at early times, we can reject the decaying solution. This leaves us with the exponentially increasing de Sitter expansion of the scale factor. It is driven by the ω_u resonant parameter, which can be expressed in terms of Λ via Eq. (6). It agrees with the observed value of Λ, because as found in the previous subsection, the magnitudes of η_Λ and η_u are the same.

2.2 Resonant amplitude and the validity of the weak-field approximation

According to Euclidian field theory, the oscillating QFT phase factors in Euclidian spacetime become Boltzmann factors in ordinary spacetime, if the field has periodic boundary conditions. When interpreted as due to a cosmic resonance, our finding that $\eta_\Lambda \approx \eta_u$ implies the existence of a periodic boundary condition with period τ_u in conformal Euclidian time. Euclidian field theory then allows us to make the identification

$$e^{i\,\omega_u\,\tau_u/c} = e^{-\omega_u\,\eta_u} \equiv e^{-\hbar\,\omega_u/(k_B T_u)}. \tag{11}$$

It gives us the temperature T_u that is induced because the time string is bounded:

$$T_u = \frac{\hbar}{k_B\,\eta_u}. \tag{12}$$

The identical result can be obtained with the help of Heisenberg's uncertainty principle. For a system in thermal equilibrium at temperature T, the equipartition theorem tells us that each degree of freedom has energy $\frac{1}{2}k_BT$. We may therefore make the identifications

$$\begin{aligned}\Delta E &\equiv \frac{1}{2}k_BT_u,\\ \Delta t &\equiv \eta_u.\end{aligned} \tag{13}$$

Inserting the value for T_u from Eq. (12) then gives us the Heisenberg relation

$$\Delta E\,\Delta t = \frac{1}{2}\hbar. \tag{14}$$

Alternatively we could have started from Eqs. (13) and (14) to obtain Eq. (12).

Replacing η_u with the Planck time t_P, we obtain the Planck temperature $T_u = T_P$. Using the definitions for the Planck time and mass,

$$\begin{aligned}t_P &= \left(\frac{\hbar G}{c^5}\right)^{1/2} \approx 5.39\times 10^{-44}\,\mathrm{s},\\ m_P &= \left(\frac{\hbar c}{G}\right)^{1/2} \approx 21.8\,\mu\mathrm{g},\end{aligned} \tag{15}$$

it follows from Eq. (12) that

$$T_P = m_Pc^2/k_B \approx 1.42\times 10^{32}\,\mathrm{K}. \tag{16}$$

This comparison serves to demonstrate that the temperature T_u and the mode energy $\hbar\omega_u$ both scale with $1/\eta_u$ throughout cosmic time all the way back to the Planck era. In Planck units the present age η_u of the universe is approximately 10^{61}, which implies that the present value of T_u is only about 10^{-29} K. Energetically this is completely insignificant in comparison with the CMB temperature. The present mode energy $\hbar\omega_u$, which is about 10^{-61} in Planck units, represents the relative amplitude by which the metric is disturbed. As long as it is $\ll 1$, one may use the Newtonian limit to interpret it as a potential energy and is allowed to use the weak-field approximation to describe it. The scaling shows that it remains $\ll 1$ everywhere, except in the nonlinear regime in the immediate vicinity of the Planck era. This tells us that the weak-field approximation is valid for all times later than about 10^{-41} s (when the amplitude was about 0.005).

2.3 Nature of the global constraint for Λ

We have shown how Λ emerges as a result of a boundary condition that exists because time in the observable universe is bounded and have referred to it as a kind of cosmic resonance. At first glance one might think that this would be some sort of cosmic Casimir effect, because the Casimir effect is known to be due to a boundary condition that limits the oscillatory modes that can exist in the quantum vacuum. Thereby measurable forces get induced.

The nature of the boundary condition is however fundamentally different in our Λ theory. The resonances of the Casimir effect are due to Dirichlet boundary conditions, when the oscillations are clamped down at the boundaries. The size of the resonant cavity is then half a wavelength, or π, for the fundamental mode. In contrast, our Λ resonance is governed by a periodic boundary condition with period 2π. Agreement with the observed value of Λ is only possible if the bounded time string has a length that corresponds to 2π. Therefore Dirichlet boundary conditions can be ruled out on observational grounds alone.

The value of Λ is tied to the value ω_u of the cosmic resonance frequency. Since ω_u is due to a global constraint, it is a constant that applies to all of the observable universe at the given epoch. In particular this means that ω_u and Λ do not vary with redshift z, for the same kind of reason that the musical tones that emanate from a violin string are not functions of position along the string. Similarly the resonances of the wave function in atoms are represented by quantum numbers, which do not vary with position within the resonating cavity but characterize the system as a whole.

The choice of observer defines the observable universe and its age. The observer is by definition always located at redshift $z = 0$ and experiences (and therefore also defines) local, dynamic time. In contrast, nonlocal, look-back time (for $z \neq 0$) cannot be experienced by any observer. In our theory Λ varies with dynamic time, but it does not vary with look-back time or redshift z. This implies that there is a fundamental distinction between dynamic time and look-back time, in contrast to standard cosmology. In the next section we will clarify how the boundedness of time and the distinction between local and nonlocal time is a consequence of the participatory role of observers in the universe.

3. The participatory role of observers

Although Einstein's opinion on the split between past, present, and future seems to have been somewhat ambivalent, his most quoted statement on the subject is that this split is an illusion, "but a very stubborn one." He tended to regard all temporal instants along the infinite timeline as somehow already preexisting as part of a 4D map. This map contains both past and future, in spite of the fact that no observer is able to directly experience any other time than what we refer to as "Now." Nevertheless the physical meaning of the concept of "Now" remained elusive to him.

The Einsteinian view of a 4D spacetime that maps all times is meaningful only in a universe devoid of observers. As soon as one introduces an observer, the timeline automatically splits up, because the presence of an observer implies a "Here" and "Now." This split is profoundly physical, because we know from experience that the future is not part of the observable universe. It is not accessible to any observer, even in principle. This is the only universe in which our cosmological theories can be tested, not in some idealized universe devoid of observers, to which nobody can belong.

We are not merely dealing with an alternative philosophical viewpoint, because the introduction of observers leads to a different physical theory with different testable consequences. In the observable universe, time is always bounded, between the Big Bang as one edge and the Now as the other edge. In contrast, in the Einsteinian universe, time is unbounded in the future. The finite temporal dimension allows a global boundary constraint that leads to the emergence of a Λ term in Einstein's equations. It is the cause of the observed acceleration of the cosmic expansion.

A fundamental difference between classical and quantum physics concerns the role of observers. We can introduce test observers in classical physics, but they are not participatory in the way that they are in quantum physics. The classical world represents an objective reality that exists in a form that is independent of the presence of observers. It is the Einsteinian universe. In contrast, the quantum reality comes into existence through the participation of observers. It is the reason for the fundamental quantum fuzziness or uncertainty, the probabilistic causality, and the irreversibility through the collapse of the wave function. While the evolution of the wave function is time symmetric and deterministic, the act of "observation" or "measurement" leads to the profoundly different nature of quantum reality.

Although the role of our cosmological observers is very different from that of quantum theory, the comparison with quantum physics serves to indicate ways in which observer participation profoundly affects the nature of the theory. While abandoning the traditional classical view by allowing observer participation, we transform the theory into something that in at least this respect is closer to the nature of quantum physics. The consequence in our case is that the value of the cosmological constant gets uniquely determined in a way that leads to a very different cosmological framework.

The presence of participating observers also changes our interpretation of spacetime in a profound way by introducing a distinction between local and nonlocal time, a distinction that is absent in a universe without observers. With nonlocal time we here mean the same thing as look-back time. In contrast, dynamical time is the same as local time, because it is the only time that an observer can experience directly. The observables are redshifts, apparent brightnesses, structuring of celestial objects, etc. The observer is by definition always at redshift $z = 0$. With the help of a cosmological model, the observables may in principle be used to infer a look-back time, which represents the way that the spacetime map appears from the vantage point of the observer.

In both standard cosmology and our alternative theory, the value of Λ applies to the totality of the observable universe at the given epoch and is therefore independent of redshift. In standard theory it is also independent of epoch (age) of the universe, while in our alternative theory, it is proportional to $1/\eta_u^2$, where η_u is the conformal age. This implies a different mathematical framework for the new cosmology, which will be developed in the next section.

4. Derivation of the cosmological evolution

The choice of observer defines the age $t = t_u$ of the universe. At proper time t_u, the scale factor is $a_u = a(t_u)$, and the Hubble constant $H = \dot{a}/a$ is $H_u = H(t_u)$. In standard cosmology the evolution of the scale factor $a(t)$, which defines the cosmological model, can be deduced from the observed relation between the expansion rate $\dot{z}/(1+z)$ (Hubble constant) and the redshift z. It is then sufficient to only consider the presently observable universe. In contrast, this is not sufficient in the nonstandard cosmology that will be developed here and which we will refer to as the alternative cosmology (AC) theory in the following. The presence of the global constraint causes the nonlocal time scale (the "look-back" time when $z > 0$) to be different from the local time scale.

The dynamical time scale is the local time scale that is experienced by a comoving observer and which characterizes the age t_u of the universe. To make this distinction clear, we attach index u to all local ($z = 0$) quantities to link them to epoch t_u, i.e., to define which observable universe they refer to.

In both standard cosmology and AC theory, the expansion rate of the universe, as represented by the Hubble constant, is governed by the equation

$$H = H_u E_u(z). \tag{17}$$

z is the redshift, and

$$E_u(z) = \left[\Omega_M(a_u)(1+z)^3 + \Omega_R(a_u)(1+z)^4 + \Omega_\Lambda(a_u)\right]^{1/2} \tag{18}$$

if we assume zero spatial curvature (see Section 4.2 for a justification of this assumption). Since H_u is defined as the local Hubble constant (at $z = 0$), it follows that $\Omega_M + \Omega_R + \Omega_\Lambda = 1$, as required for flatness. $\Omega_{M,R,\Lambda}(a_u)$ represent, respectively, the matter density (including dark matter), radiation energy density, and the "dark energy" density due to the cosmological constant Λ, all in units of the critical mass-energy density. Their values in Eq. (18) refer to the epoch when the scale factor is a_u. The relation between Ω_Λ and Λ is given by

$$\Omega_\Lambda(a_u) = \frac{c^2}{3H_u^2}\Lambda_u \tag{19}$$

as follows from Eqs. (7) and (8). In standard cosmology Λ does not depend on a_u, but in AC theory it does.

The scale factor normalized to epoch t_u is

$$y \equiv a/a_u = 1/(1+z). \tag{20}$$

The redshifts z only have a physical meaning when they refer to an epoch t_u (because this epoch is by definition where the observer at $z = 0$ exists). In terms of parameter y, the function E in Eq. (18) becomes

$$E_u(y) = \left[\Omega_M(a_u)y^{-3} + \Omega_R(a_u)y^{-4} + \Omega_\Lambda(a_u)\right]^{1/2}, \tag{21}$$

which satisfies the requirement of Eq. (17) that $E_u = 1$ when $y = 1$ or $z = 0$.

4.1 Key difference between standard cosmology and AC theory

The values of $\Omega_{M,R,\Lambda}$ that refer to the present epoch ($t_u = t_0$) can be determined by observations. In standard cosmology the parameter Λ is a true constant, independent of both redshift and epoch t_u for all times. Eq. (17) then represents a differential equation that determines the complete evolution $a(t)$ of the scale factor, when the current values of $\Omega_{M,R,\Lambda}$ are known. In contrast, in AC theory the magnitude of Λ varies with dynamical time, tracking the radius $r_u = c\eta_u$ of the causal or particle horizon. The tracking property is governed by

$$\Lambda_u = 2(\pi/r_u)^2 \tag{22}$$

according to Eq. (6). It is the fundamental equation that sets AC theory apart from standard theory.

Because the conformal age η_u is given by an integral over all times, Ω_Λ in Eq. (18) is governed by a global integral condition in AC theory. This means that the evolution of the scale factor $a(t)$ is obtained from the solution of an integrodifferential equation. With Eqs. (19) and (22), the relation between Ω_Λ and the conformal age can be expressed in the form

$$\Omega_\Lambda(\eta_u) = \frac{2}{3}\left(\frac{\pi}{x_u}\right)^2. \tag{23}$$

Here the dimensionless parameter x_u is the conformal age in units of the Hubble time $1/H_u$ at the same epoch:

$$x_u \equiv \eta_u H_u. \tag{24}$$

4.2 Theoretical justification for the flatness assumption

While observations support our assumption of vanishing spatial curvature, AC theory requires it on theoretical grounds, in contrast to standard theory. Since curvature is induced by the presence of matter-energy sources, which may include the vacuum energy ρ_Λ from a cosmological constant, an empty universe without any such sources must have zero spatial curvature. In AC theory not only the matter and radiation energy densities go to zero in the distant future but also the energy density due to the Λ term. It vanishes when the horizon radius r_u goes to infinity according to Eq. (22). At temporal infinity the universe is therefore empty, which implies flatness. When the curvature vanishes at a temporal boundary, it will remain zero for all other epochs. In contrast, in standard cosmology the density of "dark energy" (as represented by Λ) never vanishes but dominates the future dynamics. As the universe therefore never will be empty, the curvature is not constrained to be zero.

4.3 Iterative solution of the basic equations

Because the value of the conformal age η_u depends on the value of Ω_Λ, Eq. (23) is coupled to Eq. (17) in a way that most conveniently gets solved by a straightforward iteration procedure. It is found to deliver a unique value of Ω_Λ for any given value of η_u or scale factor a_u, without numerical complications. In particular, the solution for the present epoch is $\Omega_\Lambda = 67.2\%$, which is within 2σ from the value $68.5 \pm 0.7\%$ that has recently been derived from observational data with the Planck satellite [9]. It would be strange if this remarkable agreement, obtained without the use of any free parameters, would merely be a fortuitous "coincidence."

From the relation $H = \dot{a}/a$ and Eq. (17), we obtain the conformal and proper ages η_u and t_u. For convenience we express them in terms of the dimensionless functions x_u (which was already introduced in Eqs. (23) and (24)) and g_u through normalization with the Hubble time $1/H_u$:

$$\begin{aligned} x_u &\equiv \eta_u H_u = H_u \int_0^{t_u} \frac{\mathrm{d}t}{(a/a_u)} = \int_0^1 \frac{\mathrm{d}y}{y^2 E_u(y)}, \\ g_u &\equiv t_u H_u = H_u \int_0^{t_u} \mathrm{d}t = H_u \int \frac{\mathrm{d}a}{aH} = \int_0^1 \frac{\mathrm{d}y}{y E_u(y)}. \end{aligned} \tag{25}$$

To find the x_u that is needed to determine Ω_Λ via the global constraint of Eq. (23), we need to know the correct $E_u(y)$ function to be used in Eq. (25). This function however depends on the value of Ω_Λ that we want to determine. The solution can readily be obtained by iteration as follows: (i) Assume a starting value for Ω_Λ, which allows $E_u(y)$ to be defined (as clarified below). (ii) Use this $E_u(y)$ function to solve Eq. (25) for x_u, which can be inserted in Eq. (23) to obtain a new value for Ω_Λ. Insert the result in step (i) as the new starting value, and repeat the procedure until convergence. This simple iteration procedure does not encounter any numerical problems and converges quickly.

The E_u function that is used in this iteration depends not only on the value of Ω_Λ but also on the values of $\Omega_{M,R}(a_u)$ for the chosen epoch, which we define in terms of the value of the scale factor $a_u = a(t_u)$. The starting values of $\Omega_{M,R}(a_u)$ for the iteration depend on the starting value for Ω_Λ, because the flatness condition $\Omega_M(a_u) + \Omega_R(a_u) + \Omega_\Lambda(a_u) = 1$ has to be satisfied. Let us next outline how these starting values are determined.

First of all, the value of Ω_R for the radiation energy density is directly constrained by observations, because its value at the present epoch ($a_u = a_0 = 1$) is fixed by the observed values of the CMB temperature and the Hubble constant H_0. With the assumption of zero spatial curvature, the present value of Ω_M then follows from the value of Ω_Λ, because $\Omega_M = 1 - (\Omega_R + \Omega_\Lambda)$.

Ω_R is the fraction of the critical energy density $\rho_c c^2$ that is in the form of radiation energy u_R (due to photons and neutrinos):

$$\Omega_R = \frac{u_R}{\rho_c c^2}, \tag{26}$$

where

$$u_R = a_T T^4 \left[1 + \frac{7}{8}\left(\frac{4}{11}\right)^{4/3} N_\nu\right] = 1.681\, a_T T^4 \tag{27}$$

cf. [13]. $N_\nu = 3$ is the number of neutrino families, while $T = 2.725$ K is the measured temperature of the cosmic microwave background, and a_T is Stefan's constant.

When going to a different epoch with a different a_u, we change the a/a_u normalization in Eq. (20) for the scale factors a and the associated redshift scale z. Then the values of Ω_M and Ω_R must also change, because they refer to $z = 0$. Since $\Omega_M \rho_c \sim a_u^{-3}$ while $\Omega_R \rho_c \sim a_u^{-4}$, the ratio Ω_R/Ω_M scales as $1/a_u$.

During the iteration we enforce the correct a_u scaling of the $\Omega_R(a_u)/\Omega_M(a_u)$ ratio and the condition for spatial flatness, which together define the correct starting values for $\Omega_M(a_u)$ and $\Omega_R(a_u)$, once a starting value for $\Omega_\Lambda(a_u)$ has been chosen. In the case of standard cosmology, there is no iteration, because the scaling of Ω_Λ relative to $\Omega_{M,R}$ is already known. For instance, the ratios $\Omega_\Lambda/\Omega_M \sim a_u^3$ and $\Omega_\Lambda/\Omega_R \sim a_u^4$ both imply that the Λ term was insignificant in the past but dominates in the future. In contrast, in AC theory the relative contribution of Λ does not change much throughout cosmic history. At epoch a_u in standard theory, $\Omega_\Lambda(a_u) = \Omega_\Lambda / \left(\Omega_M a_u^{-3} + \Omega_R a_u^{-4} + \Omega_\Lambda\right)$, where the Ωs on the right-hand side refer to their values at the present epoch ($a_u = a_0 = 1$). Similarly, for the matter density, we have $\Omega_M(a_u) = \Omega_M a_u^{-3} / \left(\Omega_M a_u^{-3} + \Omega_R a_u^{-4} + \Omega_\Lambda\right)$, and correspondingly for the radiation energy density.

Besides $\Omega_{M,R,\Lambda}(a_u)$, $E_u(y)$, and x_u, the converged iterative solution gives us g_u from Eq. (25), which is needed for the completion of the derivation of the expansion history $a(t_u)$ of the universe, as we will see below. The whole procedure is repeated for whatever set of scale factors $a_u = a(t_u)$ that we have chosen. Here we have done the calculations for equidistant steps in $\log(a_u)$ from -12 to $+4$, on a scale where $\log a_u = 0$ corresponds to the present epoch.

The scale factor a_{eq} at equipartition between the energy densities of matter and radiation is given by

$$a_{eq} = \Omega_R(a_0)/\Omega_M(a_0). \tag{28}$$

We further note that the scale factor uniquely determines the temperature of the cosmic radiation background through

$$T_u = 2.725/a_u, \tag{29}$$

which is valid back to a temperature $T_u \approx 10^9$ K. Note that T_u is defined to represent the temperature of the photons. The numerical factor (in units of K) is fixed by the observed value of the CMB temperature at $a_u = 1$. Above $T_u \approx 10^{10}$ K the scaling with a_u is the same, and T_u is identical to the neutrino temperature T_ν, but the proportionality factor is about 40% smaller. Between approximately 10^{10} and 10^9 K, the positrons annihilate with the electrons, which leads to the release of energy in the form of gamma radiation that heats the photon gas without affecting the neutrino background. This is the reason why the photon temperature T_u has since been 40% larger than that of the neutrinos. The distinction between T_u and T_ν is of relevance for Big Bang nucleosynthesis (BBN) calculations.

Because T_u scales with $1/a_u$ according to Eq. (29) and u_R scales with T_u^4 according to Eq. (27), the radiative energy density $\Omega_R(a_u)\rho_c(a_u)$ scales with $1/a_u^4$ as required. By enforcing the $\Omega_R(a_u)/\Omega_M(a_u)$ ratio to scale as $1/a_u$ during the iteration, we are guaranteed to get the correct $1/a_u^3$ scaling for the energy density of matter $\Omega_M(a_u)\rho_c(a_u)$.

In **Figure 1** the parameters x_u (left panel) and g_u (right panel) have been plotted as functions of $\log a_u$ for AC theory (solid curves) and for standard cosmology (dashed curves). The left vertical dotted line in each panel marks the epoch of equipartition between matter and radiation, while the right dotted line represents the present epoch (when the scale factor is normalized to unity). Note how according to standard theory we happen to live at a special time when the x_u and g_u ratios are beginning to skyrocket. In contrast, in AC theory these ratios are constant at levels that are different when the universe is radiation and matter dominated, with a transition from one level to the other between the epoch of equipartition and the present time.

4.4 Solution for the time scale

The next step of the calculation is to use the solution for g_u to derive the functions for the epochs $t_u(a_u)$ and $\eta_u(a_u)$, the expansion rate $H_u(a_u)$, and the acceleration parameter $q_u(a_u)$.

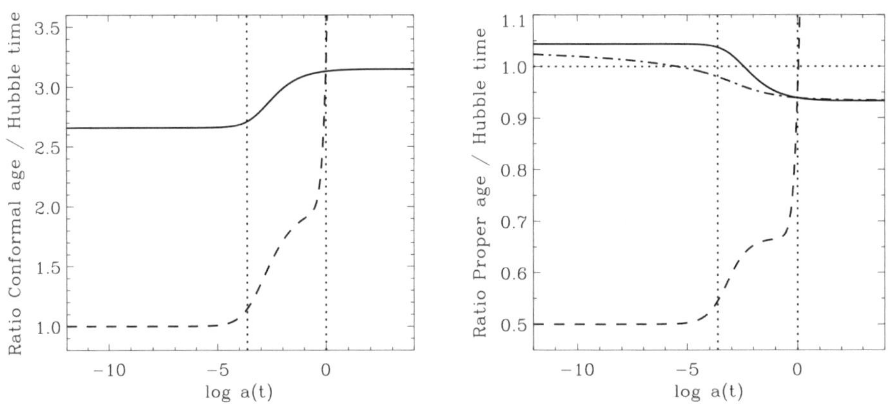

Figure 1.
The left panel shows x_u, defined by Eq. (24) as the ratio between the conformal age η_u and the Hubble time $1/H_u$, plotted vs. log of the scale factor a(t), while the right panel gives the corresponding plot for g_u, which is defined by Eq. (25) as the ratio between the proper age t_u and the Hubble time $1/H_u$. In both panels the AC theory is represented by the solid curve, the standard theory by the dashed curve. The two vertical dotted lines mark the epochs of equipartition and our present time. The dash-dotted curve in the right panel represents the exponent α in the power law representation of the scale factor in Eq. (33). According to standard theory, the current epoch marks the beginning of an inflationary phase that will last forever.

The proper age t_u can be obtained through integration of the function g_u. First we realize that the defining equation for g_u in Eq. (25) can be expressed as

$$\frac{\mathrm{d}\log a_u}{\mathrm{d}\log t_u} = H_u t_u = g_u. \tag{30}$$

It can be solved by integration to obtain the proper age t_u of the universe as a function of scale factor a_u:

$$\log t_u = \int_0^{\log a_u} (1/g_{u'})\, \mathrm{d}(\log a_{u'}) + \log t_0. \tag{31}$$

The present age t_0 is obtained from the observed value H_0 of the Hubble constant and the value of $g_u(a_0) \equiv g_0$ for the present epoch through

$$\log t_0 = \log g_0 - \log H_0, \tag{32}$$

which readily follows from the definition of g_u in Eq. (25).

The left panel of **Figure 2** shows $\log a_u$ as a function of $\log t_u/t_0$. AC theory is represented by the solid curve, standard cosmology by the dashed curve. The horizontal dotted line marks the scale factor at equipartition between matter and radiation. The slope of the AC evolution in the log-log representation is nearly constant throughout all epochs, both in the past and the future. There is nothing special about our present epoch. In contrast, according to standard theory we happen to live at the start of an inflationary phase that will be everlasting, driven by some mysterious "dark energy."

Note also that the evolutionary time scales are quite different in the two theories. While both curves coincide at the present epoch, simply because they share the same normalization $a_u = 1$ at $t_u = t_0$, the age difference diverges as we go back in time or forward into the future.

Since the AC evolution is so close to linear in the log–log diagram, it is meaningful to represent it in the form of a power law:

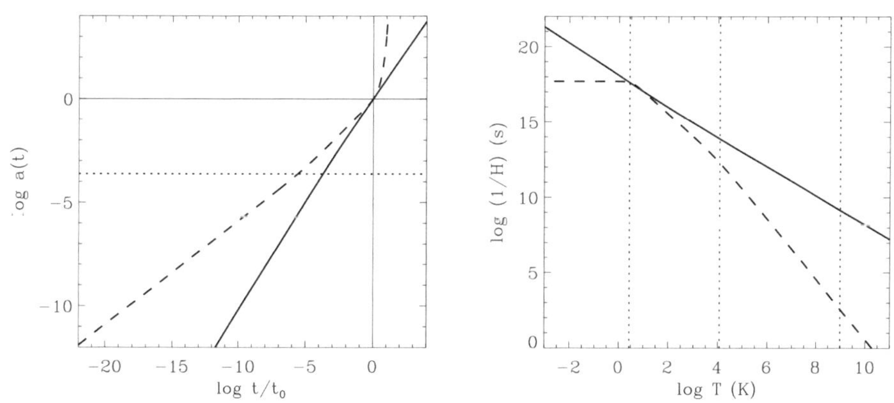

Figure 2.
In the left panel, the log of the scale factor $a(t)$ is plotted vs. log of proper time t in units of the present age t_0 of the universe. In the right panel, the log of the Hubble time in seconds is plotted vs. log of the temperature (K) of the cosmic background of electromagnetic radiation. The solid curves in both panels represent the evolution according to AC theory, while the dashed curves represent standard theory. The horizontal dotted line in the left panel marks the epoch of equipartition. The three vertical dotted lines in the right panel mark the temperatures of the present epoch, equipartition, and 1 GK (the approximate onset of nucleosynthesis). Note how in standard theory the evolution has an abrupt change at the present epoch with the onset of an inflationary phase.

$$a_u = (t_u/t_0)^{\alpha(a_u)}. \tag{33}$$

For clarity we have explicitly written the exponent α as a function of the scale factor a_u (which implies that it is also a function of time). As the function $a_u(t_u)$ is known from Eq. (31) and **Figure 2**, the functional form of α is given by

$$\alpha = \frac{\log a_u}{\log (t_u/t_0)}. \tag{34}$$

Comparison with Eq. (30) shows that α would be the same as our dimensionless function g_u if α were a true constant, independent of a_u and t_u. Since however g_u varies with a_u, the functions $\alpha(a_u)$ and $g_u(a_u)$ differ. This is illustrated in the right panel of **Figure 1**, where the function $g_u(a_u)$ as the solid curve is compared with the function $\alpha(a_u)$ as the dash-dotted curve (while g_u for standard cosmology is given by the dashed curve).

Since g_u in AC theory remains constant in the future, it coincides with the α function there, as expected. However, as we go back in time, there is a transition of g_u to a higher level, which is reached around the time of equipartition. Because of this variation, the α function initially diverges from g_u but approaches it again asymptotically as we go to ever earlier times.

Overall the temporal variations of α and g_u are very modest in AC theory, as expected from the nearly linear behavior in the left panel of **Figure 2**. In contrast, the variations are quite dramatic in standard cosmology, according to which a veritable "explosion" occurs at the present epoch, when the universe takes off in an exponential, inflationary phase.

Note that the level $\alpha = 1$, which is marked by a horizontal dotted line in the right panel of **Figure 1**, corresponds to a linear a_u vs. t_u relation with zero acceleration. Below this level we have deceleration, above it acceleration. The circumstance that g_u and $\alpha(a_u)$ in AC theory remain larger than unity in most of the radiation-dominated era of the early universe implies that the universe evolved with an accelerated expansion throughout this time. This mirrors the behavior of the acceleration parameter q_u, which will be derived and displayed in Section 5 and **Figure 3**.

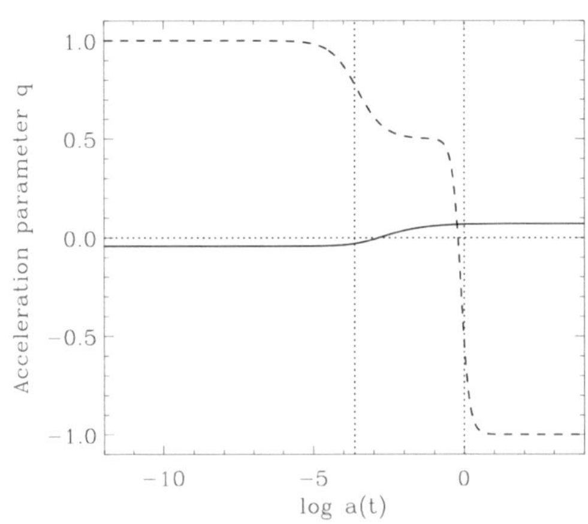

Figure 3.
Cosmic acceleration parameter q_u vs. $\log a_u$ for AC theory (solid) and standard cosmology (dashed). The vertical dotted lines mark the epochs of matter-radiation equipartition and the present time. Positive values of q imply deceleration, negative values acceleration (inflation). Note that according to AC theory the universe has been accelerating throughout the entire radiation-dominated era.

4.5 Solution for the expansion rate

Similar to Eq. (32) we obtain from the definition of g_u in Eq. (25) the expansion rate H_u as a function of $\log t_u$:

$$\log H_u = \log g_u - \log t_u. \tag{35}$$

Alternatively we may replace time t_u by the scale factor a_u via the power law description of Eq. (33), to obtain the same result in the form

$$\log H_u = -\frac{1}{\alpha(a_u)} \log a_u + \log g_u - \log t_0. \tag{36}$$

Knowing both H_u and x_u, we then get the conformal age η_u of the universe as a function of $\log a_u$ directly from Eq. (24).

In the right panel of **Figure 2**, the Hubble time $1/H_u$ is plotted vs. $\log T_u$ for AC theory (solid curve) and standard cosmology (dashed curve). The three vertical dotted lines represent, from left to right: the present epoch (T = 2.725 K), the epoch of equipartition between matter and radiation, and the BBN epoch when the radiation temperature is 10^9 K. This is the approximate temperature below which photodissociation of deuterium no longer stands in the way for Big Bang nucleosynthesis. Note how the standard theory curve has an abrupt bend at our present epoch because of the onset of an inflationary expansion. In contrast the AC theory curve remains nearly linear for all epochs, with nothing particular happening at the present epoch.

While the solid and dashed curves for $1/H_u$ coincide at the present epoch, because they obey the same observational constraint, the standard theory curve immediately diverges from the AC curve in both the future and past directions. In standard theory the expansion rate will be much faster in the future and was also much faster in the past, as compared with AC theory. This expresses the same property that was seen in the left panel of **Figure 2** for the evolution of the scale factor a. When the temperature was 10^9 K, around the BBN epoch, the age of the universe was 158 s or 2.6 min in standard cosmology, while it was 43.5 yr in AC theory, a difference by a factor of 10^7. Instead of referring to "the first 3 minutes" as the time relevant for the formation of the light elements, we would in AC theory need to refer to "the first century!"

This huge difference has major implications for our understanding of BBN physics. At a first glance, it might seem that it would make AC theory incompatible with the constraints imposed by the observed abundances of the light chemical elements, because the BBN predictions depend on the value of the expansion rate. However, a closer look at the BBN problem shows that the situation is much more complex, because we are in a totally different regime. AC theory may still be compatible with the observational constraints, but this remains an open question. At the time of writing, the required BBN modeling with AC theory is still work in progress.

Similarly the significantly slower expansion rate in AC theory around the epochs of equipartition and recombination will require a reevaluation of the processes that govern the formation of the CMB spectrum. This is needed to allow AC theory to be confronted with the constraints that are imposed by the observed CMB signatures.

5. Natural inflation without new fields

Let us next determine the cosmic acceleration parameter q_u in AC theory. The first step is to extract the time derivative $\dot{a}_u$ of the scale factor from the Hubble constant H_u:

$$\log \dot{a}_u = \log a_u + \log H_u. \tag{37}$$

Its derivative with respect to $\log t_u$ then gives us the acceleration parameter

$$q_u \equiv -\frac{\ddot{a}_u a_u}{\dot{a}_u^{\,2}} = -\frac{1}{g_u}\frac{\mathrm{d}\log \dot{a}_u}{\mathrm{d}\log t_u}. \tag{38}$$

In contrast, in standard cosmology q is obtained directly (without any use of index u) as a function of scale factor a via the relation $q(a) = [H_0/H(a)]^2$ $(0.5\Omega_M a^{-3} + \Omega_R a^{-4} - \Omega_\Lambda)$.

Figure 3 shows how standard theory (represented by the dashed curve) has three distinct levels for q. (i) In the early universe, when the universe is radiation dominated, $q = 1$. (ii) After equipartition, there is a transition to a new level ($q = 1/2$), when the universe is matter dominated. (iii) At the present epoch, there is a rapid transition to an inflationary phase that is driven by the dominating "dark energy," at the level $q = -1$.

In AC theory there is only a gentle transition around equipartition from a level of $q_u = -0.042$ when radiation dominates over matter, to a level of $q_u = +0.071$ when matter dominates over radiation. The negative level of q implies that the cosmic expansion was in an accelerated phase from the beginning of the Big Bang throughout the entire radiation-dominated era. This could be concluded already from the analysis of the right panel of **Figure 1** for g_u and $\alpha(a_u)$. Although this accelerated expansion represents a very mild form of inflation, its inflationary effect is nevertheless large, because it persists and accumulates over such a long period. It thereby accomplishes what the postulated violent inflation in the brief GUT era does. In AC theory the radiation era inflation is not postulated. Its magnitude is not a free fitting parameter but a consequence of the global resonance condition, which is the origin of the cosmological constant.

It may seem confusing that the universe is currently accelerating according to standard cosmology, while both **Figure 3** and the right panel of **Figure 1** show it to be decelerating according to AC theory. The reason is that q in standard theory refers to an apparent, nonlocal acceleration, while q_u in AC theory is the physically relevant local acceleration of the scale factor $a_u(t_u)$ in terms of the t_u time scale. This scale is different from the look-back time scale, because Λ varies with t_u.

When we throughout this chapter have referred to the "observed acceleration" of the cosmic expansion, we have implicitly meant the acceleration that is inferred when the observational data are interpreted with the Friedmann-Lemaître models, because no other framework has been available for describing the observations in terms of an evolving scale factor. The discovery with the supernova observations was that a positive cosmological constant Λ is needed to interpret the data in terms of the standard model and its magnitude could be inferred. Within this framework the inferred value of Λ means that the expansion is accelerating. However, the identical observational data with the same current value for Λ do not imply an acceleration of the local $a_u(t_u)$ scale factor within the AC theory framework. Instead the consistently derived $a_u(t_u)$ function implies a deceleration at the present epoch.

The inference of an acceleration from redshift data depends on the way in which redshift z scales with look-back time as governed by Eq. (17). AC theory does not provide any alternative definition of "look-back time." It instead explains that the physically relevant time scale is that of t_u, the age of the observable universe. This scale represents the local, dynamic time scale that is experienced by an observer (who is always located at $z = 0$). The cosmological constant in AC theory depends on t_u in contrast to standard cosmology, while being independent of redshift in both theories. Therefore the local acceleration of $a_u(t_u)$ cannot be derived from redshift observations. In summary: the acceleration inferred from supernova observations is an apparent acceleration. The physically relevant acceleration is the one that refers to the local, dynamic time scale, which is obtained with AC theory.

In standard cosmology an inflationary period in the early universe has been postulated to provide a solution to two fundamental cosmological problems: the horizon and the flatness problem [14]. The remarkable smoothness of the observed CMB tells us that the universe was homogeneous and isotropic on large scales to an extremely high degree (of order 10^{-5}) at the time of decoupling ($z \approx 10^3$). Unless one assumes extraordinarily special and improbable initial conditions, such a smoothness and isotropy can only happen, if regions in the CMB with large angular separation have been in causal contact, to allow them to interact and homogenize.

In a decelerating universe, the radius of the cosmic horizon (e.g., the Hubble radius c/H) increases faster than the expansion of space. If the dynamics of the universe were exclusively governed by matter and radiation, as in the Friedmann models without any cosmological constant, then the universe would always be decelerating. A convenient way to describe this is in terms of the "comoving Hubble radius" r_H, defined as $c/(aH)$. The temporal derivative of aH equals the acceleration of the scale factor: $\mathrm{d}(aH)/\mathrm{d}t = \mathrm{d}^2a/\mathrm{d}t^2$. This means that when the acceleration is negative (deceleration), aH decreases, and therefore its inverse (the comoving Hubble radius) increases, and vice versa if we reverse the signs.

The described properties are illustrated in **Figure 4**, where we have plotted the comoving Hubble radius r_H as a function of log of the scale factor for standard cosmology (dashed curves) and AC theory (solid curves). Before the present epoch (marked by the dotted line), r_H in the standard theory increases steeply, as a consequence of the gravitational deceleration. This deceleration is caused by the gravitational force from the radiation energy before the epoch of equipartition (marked by the vertical dash-dotted line) and by matter afterwards. Therefore the slope of the dashed curve changes around equipartition. Near the present epoch, the negative pressure from the cosmological constant begins to dominate, which marks the beginning of a phase of eternal acceleration. This causes the dashed curve to abruptly turn over and decrease steeply. Along the entire cosmic timeline, the present epoch is singled out as the epoch when this abrupt turnover takes place.

Causal contact is only possible over distances that are smaller than the comoving horizon size. As seen by the dashed curve in **Figure 4**, the largest scales that we observe today (of order 10 Gly, the approximate present Hubble radius) only came into causal contact very recently, well after the time of recombination ($a \approx 10^{-3}$). It means that they did not have time to interact and thermalize on the Hubble time scale. This makes it a mystery to standard theory why there are such strong correlations in the CMB over regions on the sky with wide angular separations.

To solve this problem, an early inflationary phase without known physical origin was postulated [14]. With its negative slope for r_H, it had the purpose of balancing out all the enhancements of r_H that have accumulated during the decelerating history of the Friedmann-type evolution of the universe, between the end of inflation until the present time. To avoid wrecking the successful BBN predictions of the Friedmann model, it was believed that the inflationary phase had to end well before the BBN era, which in standard cosmology occurs when the age of the universe is of order minutes. Inspired by the grand unification theory endeavors in particle physics, the inflationary phase is generally postulated to occur in the era of the GUT energies, when the age of the universe was somewhere between 10^{-36} and 10^{-32} s. When the dashed curve in **Figure 4** is continued back to this early era, it has decreased by many orders of magnitude, all of which must be balanced out during

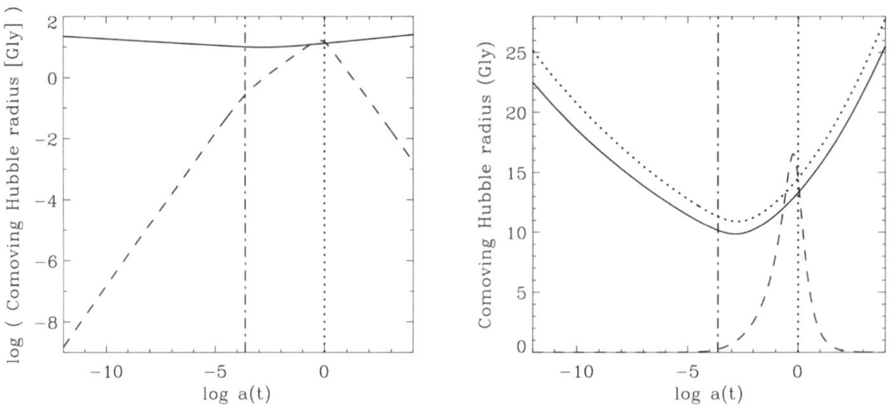

Figure 4.
Plots of the comoving Hubble radius $c/(aH)$ vs. log of the scale factor $a(t)$ for AC theory (solid curves) and standard cosmology (dashed curves). In the left panel, a log scale is used for the vertical axis, while a linear scale is used in the right panel. The vertical dash-dotted lines mark the epoch of equipartition, the dotted lines the present epoch. While H_0 based on supernova data have been used for all computations with the AC theory (and for the solid curves in this figure), the dotted curve in the right panel represents the results when H_0 from CMB data have been used instead. This serves to illustrate the degree of uncertainty that is introduced by the so-called H_0 anomaly. Note how the present epoch represents a turning point in cosmic history according to standard cosmology.

the brief inflationary phase. This is why it is generally believed that an incredibly violent inflation must have blown up the scale factor exponentially by about 60 e-foldings, which corresponds to the gigantic number of about 10^{26}.

After the inflation idea was introduced, there have been a plethora of theoretical papers on the subject, which now has a prominent place in all modern cosmology textbooks. Still, four decades after its invention, the hypothetical inflaton field that is assumed to be responsible for the phenomenon has not been identified, in spite of an abundance of searches with string theory, supersymmetric grand unified theories, or other exotic alternatives. The existence of a violently inflationary phase around the GUT era, when the universe was a tiny fraction of a second old, is often treated as a fact, while fundamental arguments against it, like in [15, 16], are largely ignored.

In contrast, the solid curve of the AC theory in **Figure 4** shows that the comoving Hubble radius r_H has never dipped below a value of 10 Gly, which represents the largest scales that are available in our present observable universe. This can be seen more clearly in the linear representation of the right panel of **Figure 4**. We already noticed in **Figures 1** and **3** that in AC theory the cosmic acceleration occurs naturally throughout the radiation-dominated phase, with a gentle transition to a decelerating phase near the epoch of recombination. Without needing to postulate or assume anything extra, without the introduction of any free parameters, we get a very extended but gentle inflationary phase that extends all the way back to the very beginning of the universe. All scales inside the horizon at the time of recombination (and CMB formation) were always inside the horizon and were therefore causally connected since the beginning. Throughout all of cosmic history until recombination, they could interact with each other and thermalize, to establish a high degree of homogeneity and isotropy. The motivation for postulating a hypothetical GUT era violent inflation does not exist in AC theory. There is no causality problem.

In the right panel of **Figure 4**, we have let AC theory be represented by two curves. The solid curve is based on the use of a value 73.5 km s^{-1} Mpc^{-1} for H_0 (the present Hubble constant) from observations of supernovae type Ia, while the dotted curve is based on 66.9 km s^{-1} Mpc^{-1} for H_0 that has been derived from the interpretation of CMB data from the Planck satellite. These two values of H_0 differ by 9.4%, a significant discrepancy that is generally referred to as the "H_0 anomaly," cf. [17]. While this anomaly is an important issue in itself, it does not affect the topics discussed in the present work. We have therefore in all our other figures elected to base all plots for the AC theory on the supernovae H_0, including the lines that mark the location of the epoch of equipartition between matter and radiation. In contrast, all the plots for the standard theory are based on H_0 from CMB data (for reasons of self-consistency of the standard framework, because all the other parameters that define standard cosmology have been determined primarily from CMB data).

The linear representation of the right panel of **Figure 4** again highlights how the present epoch is singled out by the standard model as something extraordinarily special. The comoving Hubble radius has one single narrow peak throughout all of cosmic history, and this peak is located where we happen to live in cosmic time.

6. Conclusions

The cosmological constant Λ that was needed to model the observed accelerated expansion of the universe has generally been interpreted as representing some mysterious "dark energy." However, the interpretation that dark energy is some kind of new physical field that pervades all of space leads to a cosmology (which is generally referred to as the "standard model"), in which our time in cosmic history is extraordinarily special and marks the onset of an inflationary phase that will continue forever.

Forty years ago another inflationary phase was postulated to occur in the GUT era of the very early universe, in order to answer the question why the universe is observed to be so homogeneous and isotropic on large scales [14]. The scalar inflaton field needed to drive the inflation has however not been identified in spite of a profusion of papers on this topic.

In the present work, we show that both these problems are connected and can be solved, if the Λ term that is responsible for the accelerated expansion is not a physical field but instead due to a global boundary constraint. This constraint induces a Λ term with a magnitude that tracks the conformal age η_u of the universe, such that $\Lambda \sim 1/\eta_u^2$. The density of dark energy therefore vanishes in the distant future. For the implementation of this idea, it is necessary to recognize the participatory role of observers in the universe, which has a profound effect on the nature of the theory.

We have derived and solved the mathematical equations that follow from this approach. It leads to a very different cosmological framework, which we refer to as the "AC theory" (AC for alternative cosmology). Some implications of this theory have been highlighted: The cosmic coincidence problem disappears, our epoch is not special in any way, and we are not privileged observers. The boundary constraint leads to an evolving scale factor that describes an accelerating, inflating phase from the beginning of the Big Bang throughout the entire radiation-dominated era. There is no need to postulate some early violent inflation driven by some hypothetical inflaton field, because the boundary constraint automatically causes the universe to inflate. The theory reproduces the observed value of Ω_Λ without the use of any free parameters. Because there is only one, unique solution, the possibility of parallel universes with other values of the cosmological constant does not exist.

As the cosmic expansion rate is found to have been much slower in the past than it was according to standard cosmology, the various observational data need to be reinterpreted with the new framework, in particular the BBN predictions of the abundances of the light chemical elements, and the observed signatures in the cosmic microwave background. The confrontation of the theory with such observational constraints represents work in progress that may ultimately determine the viability of the theory in its present form.

Author details

Jan Olof Stenflo[1,2]

1 Institute for Particle Physics and Astrophysics, ETH Zurich, Zurich, Switzerland

2 Istituto Ricerche Solari Locarno (IRSOL), Locarno-Monti, Switzerland

*Address all correspondence to: stenflo@astro.phys.ethz.ch

References

[1] Einstein A. Kosmologische Betrachtungen zur allgemeinen Relativitätstheorie. Berlin: Sitzungsberichte der Königlich Preußischen Akademie der Wissenschaften; 1917. pp. 142-152

[2] Riess AG, Filippenko AV, Challis P, et al. Observational evidence from supernovae for an accelerating universe and a cosmological constant. Astronomical Journal. 1998;**116**: 1009-1038

[3] Perlmutter S, Aldering G, Goldhaber G, et al. Measurements of Ω and Λ from 42 high-redshift supernovae. Astrophysical Journal. 1999;**517**:565-586

[4] Binétruy P. Dark energy and fundamental physics. Astronomy & Astrophysics Review. 2013;**21**:67

[5] Öztas AM, Dil E, Smith ML. The varying cosmological constant: A new approximation to the Friedmann equations and universe model. Monthly Notices of the Royal Astronomical Society. 2018;**476**(1):451-458

[6] Sabulsky DO, Dutta I, Hinds EA, et al. Experiment to detect dark energy forces using atom interferometry. Physical Review Letters. 2019;**123**(6): 061102

[7] Colin J, Mohayaee R, Rameez M, Sarkar S. Evidence for anisotropy of cosmic acceleration. Astronomy & Astrophysics. 2019;**631**:L13

[8] Stenflo JO. Origin of the cosmological constant. Astrophysics & Space Science. 2019;**364**(9):143

[9] Planck Collaboration, Aghanim N, Akrami Y, Ashdown M, et al. Planck 2018 results. VI. Cosmological parameters. 2018. arXiv e-prints: http://arxiv.org/abs/1807.06209v1

[10] McCoy BM. The connection between statistical mechanics and quantum field theory. 1994. arXiv e-prints: https://arxiv.org/abs/hep-th/9403084

[11] Zee A. Quantum Field Theory in a Nutshell. 2nd ed. Princeton: Princeton University Press; 2010

[12] Weinberg S. Gravitation and Cosmology: Principles and Applications of the General Theory of Relativity. New York: John Wiley & Sons, Inc.; 1972

[13] Peebles PJE. Principles of Physical Cosmology. Princeton: Princeton University Press; 1993

[14] Guth AH. Inflationary universe: A possible solution to the horizon and flatness problems. Physical Review D. 1981;**23**(2):347-356

[15] Penrose R. The Road to Reality : A Complete Guide to the Laws of the Universe. London: Vintage Books; 2004

[16] Penrose R. Fashion, Faith, and Fantasy in the New Physics of the Universe. Princeton: Princeton University Press; 2016

[17] Riess AG, Casertano S, Yuan W, et al. New parallaxes of galactic Cepheids from spatially scanning the Hubble space telescope: Implications for the Hubble constant. Astrophysical Journal. 2018;**855**:136

Chapter 5

Primary Role of the Quantum Electromagnetic Vacuum in Gravitation and Cosmology

Constantin Meis

Abstract

The electromagnetic field ground state, a zero-energy vacuum component that issues naturally from Maxwell's theory and from the vector potential quantization at a single-photon level, overcomes the vacuum energy singularity in quantum electrodynamics which leads inevitably to the well-known "vacuum catastrophe" in cosmology. Photons/electromagnetic waves are oscillations of this vacuum field which is composed of a real electric potential permeating all of space. The Hawking-Unruh temperature for a particle accelerated in vacuum is readily obtained from the interaction with the electromagnetic field ground state. The elementary charge and the electron and proton mass are expressed precisely through the electromagnetic field ground state quantized amplitude entailing that photons, leptons/antileptons, and probably baryons/antibaryons originate from the same vacuum field. Fluctuations of the electromagnetic field ground state contribute to the cosmic electromagnetic background and may be at the origin of the dark energy which is considered to be responsible for the observed cosmic acceleration. Furthermore, the gravitational constant is also expressed through the electromagnetic field ground state quantized amplitude revealing the electromagnetic nature of gravity. The overall developments yield that the electromagnetic field ground state plays a primary role in gravitation and cosmology opening new perspectives for further investigations.

Keywords: vector potential quantization, zero-point energy singularity, vacuum catastrophe, cosmological constant, electromagnetic vacuum, photon-electron-positron relation, elementary charge, mass-charge relation, electromagnetic gravity, gravitational constant

1. Introduction

Following a large number of astrophysical observations, it is actually well-established that the cosmic expansion is accelerating. This conflicts with the fundamental predictions of general relativity according to which the universe should decelerate [1–6]. The most plausible physical explanation is the cosmological constant Λ which is identified as the quantum vacuum energy density. Recent studies generally consider the dark energy to be composed mainly of the vacuum energy [4, 5, 7–9]. The cosmic acceleration has been confirmed by multiple independent studies based on different observation methods such as Type Ia supernovae (SN) [1–3, 10–13], cosmic microwave background (CMB) anisotropies [5, 14–20], weak gravitational lensing [21–24], baryon acoustic oscillations (BAO) [25–29], galaxy clusters [30–32], gamma-ray bursts [33, 34], and Hubble parameter measurements [35, 36]. Hence, there is almost no doubt today that a cosmic field with low energy density and negative pressure may provide a satisfactory explanation to the observed accelerated expansion of the universe [5, 7]. However, the identification of the cosmological constant to the vacuum energy issued from the quantum field theory leads to a serious problem related to the energy scale [4, 5, 37–39], the origin of which we analyze briefly here.

The quantization process in quantum field theory following the harmonic oscillator representation leads to the well-known puzzling singularity of infinite zero-point energy (ZPE) [40, 41] corresponding to the vacuum energy. In the case of the electromagnetic field, for example [37, 38, 42–45], in a given volume V, the ZPE density is expressed in quantum electrodynamics (QED) by the well-known relation $\rho_{ZPE} = \frac{1}{V}\sum_{k,\lambda}\frac{1}{2}\hbar\omega_k$ where $\hbar$ is Planck's reduced constant and the summation runs over all possible angular frequencies ω_k and circular polarizations λ (right and left). Transforming the discrete summation into a continuous one, according to the density of state theory [42, 44], the ZPE density becomes $\rho_{ZPE} = \frac{\hbar}{2\pi^2 c^3}\int_0^\infty \omega^3 d\omega$ which is infinite at any point in space [41]. The frequency corresponding to Planck's energy of 10^{19} GeV [5, 37, 38], that is, roughly $\sim 10^{43}$ Hz, may reasonably assumed to be a physical cutoff for the upper limit of the integration. In this case, the theoretical value obtained for the ZPE density of the electromagnetic field is around 10^{110} J m^{-3}. When considering the quantization of all other known fields, the energy scale does not radically change even if the last value gets somehow higher [4, 5, 9, 37].

On the experimental front, following the well-validated astrophysical observations mentioned above, we have good evidence today that the vacuum energy density should be approximately 10^{-9} J m^{-3}. The discrepancy between the experimental value and the different theoretical estimations is 10^{120}, the worst ever observed in science. Not surprisingly, the problem related to the quantum vacuum energy scale has been called "vacuum catastrophe" and constitutes a major challenge in modern physics [5, 37–39].

The most elaborated theoretical models on the dark energy developed up to now [45–64] are unable to resolve satisfactorily the energy scale problem. Hence, new models based on modified gravity have been advanced [65–67] obtaining interesting results although many scientists were skeptical since the beginning regarding the physical validity of such a hypothesis. Indeed, recent studies [68] of over 193 high-quality disk galaxies have finally ruled out with a high degree of statistical accuracy all modified Newtonian dynamic models. Other particular developments have been based on phenomenological assumptions [69], in particular arbitrary axioms [7], or even on the hypothesis that the physical constants like the electron charge or the fine structure constants vary with time [70] but they have not obtained any significant advancements on the problem. Finally, it is worthy to mention that the introduction of the classical notion of spin in stochastic electrodynamics (SEDS) using the real zero-point field (that is non-renormalized) yields naturally an upper frequency limit [71]. Furthermore, in this development, when approaching the upper frequency limit, the zero-point energy density is no more proportional to ω^4 but increases much slower. Consequently, SEDS has opened interesting perspectives for further studies in this field though the real energy scale problem finally remains.

The theoretical concept in QED leading to the vacuum energy singularity is based on the ZPE issued from the quantization process of the harmonic oscillator energy [40–45, 72]. It is well-known that in material harmonic oscillators, e.g., phonons in solid-state physics, the ZPE is obtained directly without any commutations of the position and momentum operators during the quantization process [41, 42, 45, 72]. Consequently, in this case a ZPE term represents a quite physical result with a direct influence on the thermodynamic properties of materials, e.g., the specific heat. Conversely, during the quantization process of the electromagnetic field, commutations between the position and momentum operators occur unavoidably leading to the "normal ordering" Hamiltonian without the ZPE and to the "anti-normal ordering" one involving a ZPE term [40–45, 72]. It has been pointed out [42, 45] that this mathematical procedure suffers from the fundamental ambiguity consisting of replacing products of naturally commuting classical canonical variables by products of non-commuting quantum mechanics operators and consequently may lead to unphysical results [42, 73]. In fact, no experiments have ever demonstrated that a single-photon state is a harmonic oscillator. Hence, in QED the "normal ordering" Hamiltonian, which is not a harmonic oscillator, is the only principally employed in all calculations dropping aside the ZPE singularity.

Regarding the vacuum effects, like the spontaneous emission and the Lamb shift, they are interpreted in QED [41, 42, 44] based on the fundamental commutation properties of the creation $a_{k\lambda}^{+}$ and annihilation operators $a_{k\lambda}$ of a k-mode and λ-polarization photon without invoking the harmonic oscillator ZPE expression. The reason is simply that the ZPE term is a constant and has absolutely no influence in the QED calculations because it commutes with all Hermitian operators $\tilde{Q}$ corresponding to physical observables $\left[\tilde{Q}, \sum_{k,\lambda} \frac{1}{2}\hbar\omega_k\right] = 0$.

Finally, due to the unobserved impact of the zero-point energy singularity in cosmology, it becomes progressively more and more accepted today that the direct interpretation of the Casimir effect based on the source fields [74, 75] or Lorentz forces [76] without invoking at all the electromagnetic field zero-point energy should be the real physical explanation of this effect [77]. In fact, from the historical point of view, the interpretations of the Casimir effect based on the ZPE had been carried out well before the astrophysical observations [1, 2, 5, 10] have ruled out the corresponding vacuum concept.

In what follows we show that the vacuum energy singularity is overcome by enhancing the vector potential amplitude quantization to a single-photon state. This procedure issues naturally from Maxwell's theory and yields a zero-energy electromagnetic field ground state capable of generating photons. The lepton/antilepton and proton/antiproton charge, the electron and proton mass, and the gravitational constant are expressed exactly through the quantized amplitude of the electromagnetic field ground state putting in evidence that it plays a fairly important role in cosmology.

2. Vector potential amplitude of a cavity-free photon

A detailed dimension analysis of the vector potential general solution obtained from Maxwell's equations shows that it is proportional to a frequency [42, 72, 78, 79]. Consequently, we may write the vector potential amplitude α_{0k} for a single free k-mode photon with angular frequency ω_k as follows [45, 80–84]:

$$\alpha_{0k}(\omega_k) = \xi\omega_k \tag{1}$$

where ξ is a constant.

It is worthy to notice that Eq. (1) is not an arbitrary hypothesis but a mathematical representation resulting directly from Maxwell's equations [45, 72]. The normalization of the energy of a single k-mode plane electromagnetic wave over a wavelength to Planck's experimental expression for the photon energy $\hbar\,\omega_k$ leads to the evaluation of the constant ξ [45, 80]:

$$|\xi| = \left|\frac{\hbar}{4\pi e c}\right| = 1.747\ 10^{-25}\ \mathrm{V\,m^{-1}s^2} \tag{2}$$

where c is the speed of light in vacuum and e is the electron/positron charge.

Eq. (2) expresses the physical relation between Planck's constant and the electromagnetic nature of the photon through the vector potential amplitude.

By this way, Eq. (1) permits to complement the fundamental physical properties relation characterizing the *wave-particle* nature of a single k-mode photon in vacuum by introducing the missing electromagnetic nature through the quantized vector potential amplitude:

$$\frac{E_k}{\hbar} = \frac{\left|\vec{p}_k\right|}{\hbar/c} = \frac{\alpha_{0k}}{|\xi|} = \left|\vec{k}\right| c = \omega_k \tag{3}$$

The last relation signifies that the particle properties of the photon, that is, energy E_k and momentum $\vec{p}_k$, and the electromagnetic wave properties, that is, vector potential amplitude α_{0k} and wave vector $\vec{k}$, are all related to the angular frequency ω_k.

Thus, the vector potential function of a free single photon can now be written in the plane wave representation [45, 80, 81]:

$$\vec{\alpha}_{k\lambda}\left(\vec{r},t\right) = \xi\omega_k\left(\hat{\varepsilon}_{k\lambda}e^{i\left(\vec{k}\cdot\vec{r}-\omega_k t+\theta\right)} + \hat{\varepsilon}^*_{k\lambda}e^{-i\left(\vec{k}\cdot\vec{r}-\omega_k t+\theta\right)}\right) = \omega_k\vec{\Xi}_{k\lambda}\left(\vec{r},t\right) \tag{4}$$

where λ denotes a circular polarization (left or right), $\hat{\varepsilon}_{k\lambda}$ is the corresponding complex unit vector, and θ is a phase parameter.

The last equation can also be written in QED representation as a function of the creation and annihilation operators $a^+_{k\lambda}$ and $a_{k\lambda}$, respectively, for a k-mode and λ-polarization photon:

$$\tilde{\alpha}_{k\lambda}\left(\vec{r},t\right) = \xi\omega_k\left(\hat{\varepsilon}_{k\lambda}a_{k\lambda}e^{i\left(\vec{k}\cdot\vec{r}-\omega_k t+\theta\right)} + \hat{\varepsilon}^*_{k\lambda}a^+_{k\lambda}e^{-i\left(\vec{k}\cdot\vec{r}-\omega_k t+\theta\right)}\right) = \omega_k\tilde{\Xi}_{k\lambda}\left(\vec{r},t\right) \tag{5}$$

Notice that the main function $\Xi_{k\lambda}\left(\vec{r},t\right)$ of the vector potential expressed in both representations constitutes the physical "skeleton" of photons/electromagnetic waves.

It is a straightforward calculation to show [45, 82, 83] that the photon vector potential function $\vec{\alpha}_{k\lambda}\left(\vec{r},t\right)$ satisfies the classical wave propagation equation in vacuum:

$$\vec{\nabla}^2\vec{\alpha}_{k\lambda}\left(\vec{r},t\right) - \frac{1}{c^2}\frac{\partial^2}{\partial t^2}\vec{\alpha}_{\kappa\lambda}\left(\vec{r},t\right) = 0 \tag{6}$$

as well as the *vector potential energy* (wave-particle) equation for the photon

$$i\left(\frac{\xi}{\hbar}\right)\frac{\partial}{\partial t}\vec{\alpha}_{k\lambda}\left(\vec{r},t\right) = \left(\frac{\tilde{\alpha}_0}{\tilde{H}}\right)\vec{\alpha}_{k\lambda}\left(\vec{r},t\right) \tag{7}$$

where $\tilde{H} = -i\hbar c\vec{\nabla}$ is the relativistic massless particle Hamiltonian having eigenvalue the single-photon energy $\hbar\omega_k$ and $\tilde{\alpha}_0 = -i\xi c\vec{\nabla}$ is the vector potential amplitude operator having eigenvalue the single-photon vector potential amplitude $\xi\omega_k$.

Eq. (7) is simply a combination of Schrödinger's equation for the energy to a symmetrical wave equation for the vector potential [45, 72, 82] expressing the simultaneous wave-particle nature of the photon.

From the operator expressions and the corresponding eigenvalues for the energy and the vector potential amplitude, we readily define an angular frequency operator $\tilde{\Omega}$ which writes

$$\tilde{\Omega} = -ic\vec{\nabla} \tag{8}$$

so that the Hamiltonian and the vector potential amplitude operators can be expressed simply as

$$\tilde{H} = \hbar\tilde{\Omega}; \quad \tilde{\alpha}_0 = \xi\tilde{\Omega} \tag{9}$$

We can thus obtain the equation governing the main function $\vec{\Xi}_{k\lambda}\left(\vec{r},t\right)$ of the vector potential in vacuum by introducing the angular frequency operator in the *vector potential-energy* Eq. (7):

$$i\frac{\partial}{\partial t}\vec{\Xi}_{k\lambda}\left(\vec{r},t\right) = \tilde{\Omega}\vec{\Xi}_{k\lambda}\left(\vec{r},t\right) \tag{10}$$

Consequently, photons/electromagnetic waves are generated by the action of the angular frequency operator $\tilde{\Omega}$ upon the fundamental function $\vec{\Xi}_{k\lambda}\left(\vec{r},t\right)$ creating a real vector potential:

$$\tilde{\Omega}\vec{\Xi}_{k\lambda}\left(\vec{r},t\right) = -ic\vec{\nabla}\vec{\Xi}_{k\lambda}\left(\vec{r},t\right) = \omega_k\vec{\Xi}_{k\lambda}\left(\vec{r},t\right) = \vec{\alpha}_{k\lambda}\left(\vec{r},t\right) \tag{11}$$

The vector potential function $\vec{\alpha}_{k\lambda}\left(\vec{r},t\right)$ expressed in Eq. (4) can be considered as a real wave function for the photon [45, 82–84]. In fact, previous attempts based on the electric and magnetic fields failed to define satisfactorily a photon wave function [85–89]. Here, the vector potential function $\vec{\alpha}_{k\lambda}\left(\vec{r},t\right)$ with the quantized amplitude $\xi\omega_k$ expresses a real probability amplitude entailing that the probability for localizing a photon is proportional to the square of the angular frequency:

$$P_k\left(\vec{r}\right) \propto \left|\vec{\alpha}_{k\lambda}\left(\vec{r},t\right)\right|^2 \propto \xi^2\omega_k^2 \tag{12}$$

This is in agreement with the experimental evidence following which the higher the frequency, the better the localization probability for a single photon [42, 44, 78].

Weighting the vector potential function by $\omega_k\sqrt{2\varepsilon_0}$ and considering both circular polarizations ($\lambda = L, R$), a six-component general wave function can be defined for the photon:

$$\Phi_{k,(L,R)}\left(\vec{r},t\right) = \omega_k\sqrt{2\varepsilon_0}\begin{pmatrix}\vec{\alpha}_{kL}\left(\vec{r},t\right)\\ \vec{\alpha}_{kR}\left(\vec{r},t\right)\end{pmatrix} \tag{13}$$

which is now suitably normalized in order to get the energy density of the electromagnetic field composed of a single k-mode

$$\left|\Phi_{k,(L,R)}\left(\vec{r},t\right)\right|^2 = 2\varepsilon_0\xi^2\omega_k^4 \tag{14}$$

From the photon vector potential, we also deduce that a single photon has intrinsic electric $\vec{\varepsilon}_k$ and magnetic $\vec{\beta}_k$ fields whose amplitudes in vacuum are proportional to the square of the angular frequency [45, 82, 83]:

$$\left|\vec{\varepsilon}_k\right| = \left|-\frac{\partial}{\partial t}\vec{\alpha}_{k\lambda}\left(\vec{r},t\right)\right| \propto \xi\omega_k^2 \text{ and } \left|\vec{\beta}_k\right| \propto \sqrt{\varepsilon_0\mu_0}\,\xi\omega_k^2 \tag{15}$$

where ε_0 and μ_0 are the vacuum electric permittivity and magnetic permeability, respectively.

Eqs. (3), (14), and (15) clearly show that all the physical properties characterizing a single k-mode photon as an integral wave-particle entity of the electromagnetic field depend directly on the angular frequency ω_k.

3. The electromagnetic field ground state as a vacuum field and the Hawking-Unruh temperature

From Eqs. (4) and (5) appears clearly that the photon vector potential is mainly composed of the fundamental field $\Xi_{k\lambda}\left(\vec{r},t\right)$. As developed in the previous section, a photon subsists only for a non-zero angular frequency ω_k characterizing the rotation (left or right) of the vector potential perpendicularly to the propagation axis generating an electric and magnetic field whose amplitudes are given in Eq. (15). Now, it is interesting to investigate what happens at zero frequency. Following Eqs. (3), (14), and (15), we can draw that the zero-frequency level ($\omega_k \rightarrow 0$) of the electromagnetic field corresponds to a cosmic state (the wavelength $\lambda_k = \frac{2\pi c}{\omega_k} \rightarrow \infty$) characterized by the complete absence of the photon physical properties: energy, energy density, vector potential, and electric and magnetic fields are all zero. This state lays beyond the Ehrenberg-Siday and Bohm-Aharonov physical situation in which the electric and magnetic fields are zero but space is filled by a real vector potential [90, 91].

However, at $\omega_k = 0$ the resulting electromagnetic field state is not synonym to perfect vacuum because the fundamental function $\Xi_{k\lambda}\left(\vec{r},t\right)$ of the vector potential gets reduced to the field $\Xi_{0\lambda}$ which writes in both representations:

$$\vec{\Xi}_{0\lambda} = \xi\left(\hat{\varepsilon}_\lambda e^{i\theta} + \hat{\varepsilon}_\lambda^* e^{-i\theta}\right); \tilde{\Xi}_{0\lambda} = \xi\left(\hat{\varepsilon}_\lambda a_{k\lambda} e^{i\theta} + \hat{\varepsilon}_\lambda^* a_{k\lambda}^+ e^{-i\theta}\right) \tag{16}$$

Electromagnetic fields are real [79, 92], and the reality of the vector potential has been well established experimentally [90, 93–95]; consequently the fundamental function $\Xi_{k\lambda}\left(\vec{r},t\right)$ in Eqs. (4) and (5) is also real. At the limit $\omega_k \rightarrow 0$ the residual field $\Xi_{0\lambda}$ is a real field permeating all of space ($\lambda_k \rightarrow \infty$) and according to Eq. (2) has an electric potential amplitude with units V m^{-1} s^2. Thus, $\Xi_{0\lambda}$ corresponds physically to the electromagnetic field ground state, a dark cosmic field capable of generating any k-mode photon with left or right circular polarization and which in absence of energy and vector potential can be considered as a vacuum component, identical in both classical electromagnetic theory and QED.

Heisenberg's energy-time uncertainty relation applied in Eq. (3) entails directly that the vector potential amplitude is also subject to a fluctuation uncertainty:

$$\delta E_k \cdot \delta t \approx \hbar \rightarrow \delta \alpha_{0k} \cdot \delta t \approx \xi \tag{17}$$

Consequently, fluctuations of the electromagnetic field ground state in space imply that transient states of various k-mode and λ-polarization photons can be generated spontaneously during time intervals respecting Heisenberg's relation contributing to the cosmic radiation background and to its associated anisotropies and might be at the origin of the dark energy. From Eq. (17), we deduce that the lifetime is longer for the low-frequency transient photons and consequently we can expect a quite important contribution in the cosmic radio background at long wavelengths. It would be extremely worthy to investigate experimentally the very low frequency cosmic radiation background spectrum.

The phase parameter θ in Eq. (16) can take any value, and consequently the electromagnetic field ground state contains all possible main functions $\Xi_{k\lambda}$ corresponding to all modes and polarizations. Hence, according to Eq. (10), any perturbation expressed through an angular frequency operator may create real photons in space. It can be easily demonstrated [45, 80] that the electromagnetic field ground state complements the normal ordering Hamiltonian representation in QED by getting a direct interpretation of the vacuum effects. Indeed, an interaction Hamiltonian between the electrons and the vacuum field $\Xi_{0\lambda}$ can be readily defined resulting precisely to the spontaneous emission rate. Also, it is important to notice that the vector potential operator in the interaction Hamiltonian used in Bethe's [96] and Kroll's [97] calculations for the Lamb effect can be replaced by that of Eq. (5) yielding exactly the same energy shifts.

We have mentioned previously that the vacuum effects, that is, the spontaneous emission and the Lamb shift, are interpreted in QED [41, 42, 44, 96, 97] without invoking the ZPE of the electromagnetic field. The Casimir effect is equally well explained [74–77] without invoking at all the ZPE which inevitably leads to the "vacuum catastrophe." Now, it can be easily demonstrated that the Hawking-Unruh temperature [98], associated to the Fulling-Davies-Unruh effect [99–101] for a charge accelerated in vacuum, can also be deduced without invoking the ZPE. In fact, any particle moving in the electromagnetic field ground state with an acceleration $\vec{\gamma}$ experiences an electric potential:

$$U = \left|\xi \vec{\gamma}\right| \tag{18}$$

Notice that even for high relativistic values, of the order of $\left|\vec{\gamma}\right| \propto 10^7 \mathrm{m\,s^{-2}}$, the electric potential felt by the accelerated particle is very low $U \propto 10^{-18} \mathrm{V}$.

For a charge e, the corresponding energy along a given degree of freedom is equivalent to a thermal energy according to the equipartition theorem:

$$E = \left|e \xi \vec{\gamma}\right| = \frac{1}{2} k_B T \tag{19}$$

where k_B is Boltzmann's constant.

Replacing ξ in the last equation by the expression of Eq. (2), one gets directly the Hawking-Unruh temperature:

$$T = \frac{\hbar}{2\pi c k_B} \left|\vec{\gamma}\right| \tag{20}$$

This extremely simple calculation shows that an accelerated charge in the electromagnetic field ground state will "feel" the Hawking-Unruh temperature.

In fact, there are many experimental controversies in the literature related to the measurement of the Hawking-Unruh temperature and to the physical reality of the Fulling-Davies-Unruh effect [102, 103]. Following the above calculation, the measure of the electric potential energy variation of the accelerated particle could be more affordable experimentally than the direct measure of such a low temperature and could consequently lead to a real validation of Eq. (19).

4. The electromagnetic field ground state, the charge-mass relation, and the gravitational constant

When replacing Planck's constant in the photon energy $E_k = \hbar\omega_k$ by an equivalent expression obtained from the fine structure constant $\alpha = e^2/4\pi\varepsilon_0\hbar c \approx 1/137$, which is dimensionless, then the energy of a free photon depends directly on the electron charge. This was always quite puzzling, and it has been often advanced [42, 43, 78] that photons and electrons/positrons should be strongly related physical entities.

Now, from Eq. (2) and the fine structure constant expression, we straightforward draw that the lepton/antilepton and the proton/antiproton elementary charge, a fundamental physical constant, is expressed exactly through the electromagnetic field ground state quantized amplitude constant ξ:

$$e = \pm(4\pi)^2\alpha\frac{|\xi|}{\mu_0} = \pm 1.602\,10^{-19}\mathrm{C} \tag{21}$$

where α is the fine structure constant, $\mu_0 = 4\pi\,10^{-7}\mathrm{H\,m^{-1}}$ is the vacuum magnetic permeability, and $|\xi| = 1.747\,10^{-25}\mathrm{V\,m^{-1}s^2}$.

The last relation shows that the single-photon vector potential and the elementary charge are related directly to the electromagnetic field ground state through the quantized amplitude constant ξ. This supports further the strong physical relationship between photons and electrons/positrons which appear to originate from the same vacuum field being consequently at the origin of their mutual transformation mechanism.

Recalling that the electron and proton mass at rest are expressed as $m_e = e\hbar/2\mu_B$ and $M_P = e\hbar/2\mu_P$, respectively, where $\mu_B = 9.274\ 10^{-24}\ \mathrm{JT^{-1}}$ is the Bohr magneton and $\mu_P = 5.0508\ 10^{-27}\ \mathrm{JT^{-1}}$ is the proton magneton, and using again Eq. (2), we deduce the relations of the electron and proton mass depending also on the constant ξ:

$$m_e = 2\pi c\,e^2\left|\frac{\xi}{\mu_B}\right| = 9.109\ 10^{-31}\mathrm{kg} \tag{22}$$

$$M_P = 2\pi c\,e^2\left|\frac{\xi}{\mu_P}\right| = 1.672\ 10^{-27}\mathrm{kg} \tag{23}$$

Notice that the ratio of the proton-to-electron mass equals the ratio of the electron-to-proton magneton $M_P/m_e = \mu_e/\mu_P = 1836.15$ according to the experimental evidence. Eqs. (22) and (23) show that the electron and proton mass is also related directly to the electromagnetic field ground state through the vector potential amplitude constant ξ yielding the quite interesting conclusion that the electron and proton mass are equally manifestations of this field and depend on the elementary charge and on the associated magnetic moments.

It has been shown [104] that the masses of all the fundamental elementary particles can be obtained from the electron mass and the fine structure constant with a precision of roughly 1%.

Consequently, the mass m_i of any elementary particle i can be expressed using Eq. (22)

$$m_i = 2\pi c\, e^2 \left| \frac{\xi}{|\mu_i|} \right| \tag{24}$$

with $|\mu_i| = \mu_B$ for the electron and $|\mu_i| = \left(\frac{2\alpha}{n_i}\right)\mu_B$ for other particles where n_i is simply an integer and α is the fine structure constant.

This formalism is valid for leptons (e.g., muon for n_i = 3, tau for n_i = 51), mesons (e.g., pion for n_i = 4, kaon for n_i = 14, rho for n_i = 22, ... etc.) as well as baryons (e.g., nucleon for n_i = 27, lambda for n_i = 32, sigma for n_i = 34, ... etc.).

A generalization of these results means that:

- charges are states of the electromagnetic field ground level,
- particle masses issue from charges and their corresponding magnetic flux; hence, all the neutral particles should be composed of positive and negative charges,
- gravitation is consequently an electromagnetic effect.

Spontaneous creation of particle/antiparticle pairs during short time-intervals due to the electromagnetic field ground state fluctuations may occur in space. We can make the hypothesis here that other type of unknown particle/antiparticle pairs could also emerge from the electromagnetic field ground state so that the overall process in the universe may contribute to the cosmic mass background and eventually to the dark matter [4, 9]. Hence, the electromagnetic field ground state appears to be a cosmic source of energy (photons) and charges (mass).

Recent observations [105, 106] have indicated that space granularity should be many orders of magnitude less than Planck's length, usually denoted as l_P and having the value of l_P = 1.616 10^{-35} m. However, Planck's length is generally considered as a characteristic physical parameter for the electromagnetic field corresponding theoretically to the shorter possible wavelength of a photon [4, 9]. This corresponds to a photon frequency close to 10^{43} Hz. Although we have not yet observed photons with such a high energy, no photon can be conceived, at least theoretically, beyond this upper frequency limit.

Therefore, we can draw now another result related to the gravitational constant G which can be expressed exactly by the square of the ratio of Planck's length l_P to the electromagnetic field ground state quantized amplitude ξ:

$$G = \frac{1}{(4\pi)^3 \alpha\, \varepsilon_0} \left(\frac{l_P}{\xi}\right)^2 = 6.674\ 10^{-11} \mathrm{m}^3\ \mathrm{kg}^{-1}\ \mathrm{s}^{-2} \tag{25}$$

where α is the fine structure constant and $\varepsilon_0 = 8.854\ 10^{-12} \mathrm{F\,m}^{-1}$ is the electric permittivity of vacuum. Introducing the complete expression of α in the last equation and taking into account Eq. (2), we deduce that the gravitational constant G, the elementary charge e, and the vector potential amplitude constant ξ are directly related as follows:

$$G = \frac{l_P^2 c^2}{4\pi e\, \xi} \tag{26}$$

According to the last equation, the electromagnetic character of gravity appears clearly entailing new possibilities for theoretical and experimental investigations in this field [107].

5. Conclusions

The vacuum concept initially identified as the zero-point energy singularity of the quantized fields has been ruled out by recent well-validated astrophysical observations. Instead, the electromagnetic field ground state $\Xi_{0\lambda}$, a zero-energy cosmic dark field permeating all of space and having the real amplitude $\xi = \hbar/4\pi ec$, issues naturally from Maxwell's theory and is compatible with the observational evidence. It is readily deduced that photons/electromagnetic waves, are oscillations of this vacuum field which is identical in classical electromagnetic wave theory and QED. Thus, the electromagnetic field ground state naturally complements the normal ordering Hamiltonian in QED overcoming the zero-point energy singularity.

Fluctuations of the electromagnetic field ground state may give birth to transient photons contributing to the observed vacuum energy density, considered responsible for the cosmic acceleration, as well as to the cosmic radiation background and to its anisotropies.

The elementary charge issues from the electromagnetic field ground state and is expressed exactly through the constant ξ. This demonstrates the strong physical relationship between photons and leptons/antileptons. The mechanisms governing their mutual transformations are directly related to the nature of the electromagnetic field ground state. Furthermore, it is shown that a charge accelerated in the electromagnetic field ground state will experience the Hawking-Unruh temperature.

Like photons, transient pairs of particles/antiparticles may emerge from the electromagnetic field ground state fluctuations contributing to the cosmic matter background and eventually to the dark matter.

It is also drawn that mass issues from charges which appear to be states of the electromagnetic field ground state revealing that the last one is a cosmic source of energy (photons) and charges (mass). Finally, the gravitational constant can be expressed exactly through the elementary charge and the electromagnetic field ground state amplitude entailing that gravitation has an electromagnetic nature and putting in evidence the primary role the electromagnetic vacuum might play in gravitation and cosmology.

Author details

Constantin Meis[1,2]

1 National Institute for Nuclear Science and Technology, CEA - Saclay, France

2 Department of Physics, University of Versailles, UVSQ, France

*Address all correspondence to: constantin.meis@cea.fr; constantin.meis@uvsq.fr

References

[1] Riess AG et al. Observational evidence from supernovae for an accelerating universe and a cosmological constant. Astronomy Journal. 1998;**116**: 1009-1038

[2] Perlmutter S et al. Measurements of Ω and Λ from 42 high redshift supernovae. The Astrophysical Journal. 1999;**517**:565-586

[3] Holanda RFL, Pereira SH. Can galaxy clusters, type Ia supernovae, and the cosmic microwave background rule out a class of modified gravity theories. Physical Review D. 2016;**94**:104037

[4] Mukhanov V. Physical Foundations of Cosmology. Cambridge, UK: Cambridge University Press; 2005

[5] Frieman JA, Turner MS, Huterer D. Dark energy and the accelerating universe. Annual Review of Astronomy and Astrophysics. 2008;**46**:385-432

[6] Linder EV. Mapping the cosmological expansion. Reports on Progress in Physics. 2008;**71**:5

[7] Beck C. Axiomatic approach to the cosmological constant. Physica A. 2009; **388**:3384-3390. arXiv:0810.0752

[8] Peebles PJE, Ratra B. The cosmological constant and dark energy. Reviews of Modern Physics. 2003;**75**:559

[9] Hey T. The New Quantum Universe. New York, U.S: Cambridge University Press; 2003

[10] Riess AG et al. Type Ia supernova discoveries at z > 1 from the Hubble space telescope: Evidence for past deceleration and constraints on dark energy evolution. The Astrophysical Journal. 2004;**607**:665-687

[11] Scolnic D et al. SUPERCAL: Cross calibration of multiple photometric systems to improve cosmological measurements with type Ia supernovae. The Astrophysical Journal. 2015; **815**(2):117

[12] Supernova Cosmology Project Collaboration, Knop RA, et al. New constraints on Omega(M), Omega (lambda), and w from an independent set of eleven high-redshift supernovae observed with HST. The Astrophysical Journal. 2003;**598**:102

[13] Riess AG et al. New Hubble space telescope discoveries of type Ia supernovae at z > 1: Narrowing constraints on the early behavior of dark energy. The Astrophysical Journal. 2007;**659**:98-121

[14] Spergel D et al. First year Wilkinson Microwave Anisotropy Probe (WMAP) observations: Determination of cosmological parameters. Astrophysical Journal Supplement. 2003;**148**:175

[15] Spergel D et al. Wilkinson Microwave Anisotropy Probe (WMAP) three year results: Implications for cosmology. Astrophysical Journal Supplement. 2007;**170**:377

[16] Komatsu E et al. Five-Year Wilkinson Microwave Anisotropy Probe (WMAP) observations: Cosmological interpretation. Astrophysical Journal Supplement. 2009;**180**:330

[17] Komatsu E et al. Seven-Year Wilkinson Microwave Anisotropy Probe (WMAP) observations: Cosmological interpretation. Astrophysical Journal Supplement. 2011;**192**:18

[18] Boomerang Collaboration, de Bernardis P, et al. A flat universe from high resolution maps of the cosmic microwave background radiation. Nature. 2000;**404**:955-959

[19] Corasaniti PS, Kunz M, Parkinson D, Copeland EJ, Bassett BA. The foundations of observing dark energy dynamics with the Wilkinson microwave anisotropy probe. Physical Review D. 2004;**70**:083006

[20] Weller J, Lewis AM. Large scale cosmic microwave background anisotropies and dark energy. Monthly Notices of the Royal Astronomical Society. 2003;**346**:987-993

[21] Crittenden RG, Natarajan P, Pen U-L, Theuns T. Discriminating weak lensing from intrinsic spin correlations using the curl-gradient decomposition. The Astrophysical Journal. 2002;**568**:20

[22] Asaba S, Hikage C, Koyama K, Zhao G-B, Hojjati A, et al. Principal component analysis of modified gravity using weak lensing and peculiar velocity measurements. JCAP. 2013;**1308**:029

[23] Zhang P, Liguori M, Bean R, Dodelson S. Probing gravity at cosmological scales by measurements which test the relationship between gravitational lensing and matter overdensity. Physical Review Letters. 2007;**99**:141302

[24] Vanderveld RA, Mortonson MJ, Hu W, Eier T. Testing dark energy paradigms with weak gravitational lensing. Physical Review D. 2012;**85**: 103518

[25] SDSS Collaboration, Eisenstein DJ, et al. Detection of the baryon acoustic peak in the large-scale correlation function of SDSS luminous red galaxies. The Astrophysical Journal. 2005;**633**: 560-574

[26] Percival WJ, Cole S, Eisenstein DJ, Nichol RC, Peacock JA, Pope AC, et al. Measuring the baryon acoustic oscillation scale using the SDSS and 2dFGRS. Monthly Notices of the Royal Astronomical Society. 2007;**381**: 1053-1066

[27] Blake C et al. The wiggle Z dark energy survey: Mapping the distance-redshift relation with baryon acoustic oscillations. Monthly Notices of the Royal Astronomical Society. 2011;**418**: 1707-1724

[28] Beutler F, Blake C, Colless M, Jones DH, Staveley-Smith L, Campbell L, et al. The 6dF galaxy survey: Baryon acoustic oscillations and the local Hubble constant. Monthly Notices of the Royal Astronomical Society. 2011;**416**:3017-3032

[29] BOSS Collaboration, Anderson L, et al. The clustering of galaxies in the SDSS-III Baryon Oscillation Spectroscopic Survey: Baryon acoustic oscillations in the Data Releases 10 and 11 Galaxy samples. Monthly Notices of the Royal Astronomical Society. 2014;**441**(1):24-62

[30] Allen SW, Evrard AE, Mantz AB. Cosmological parameters from observations of galaxy clusters. Annual Review of Astronomy and Astrophysics. 2011;**49**:409-470

[31] Basilakos S, Perivolaropoulos L. Testing GRBs as standard candles. Monthly Notices of the Royal Astronomical Society. 2008;**391**: 411-419

[32] Mantz A, Allen SW, Rapetti D, Ebeling H. The observed growth of massive galaxy clusters I: Statistical methods and cosmological constraints. Monthly Notices of the Royal Astronomical Society. 2010;**406**:1759-1772

[33] Amati L et al. Intrinsic spectra and energetics of BeppoSAX gamma-ray bursts with known redshifts. Astronomy and Astrophysics. 2002;**390**:81

[34] Wang Y. Model-independent distance measurements from gamma-ray bursts and constraints on dark energy. Physical Review D. 2008;**78**: 123532

[35] Freedman WL, Madore BF. The Hubble constant. Annual Review of Astronomy and Astrophysics. 2010;**48**: 673-710

[36] Yu H-R, Lan T, Wan H-Y, Zhang T-J, Wang B-Q. Constraints on smoothness parameter and dark energy using observational H(z) data. Research in Astronomy and Astrophysics. 2011;**11**: 125-136

[37] Adler RJ, Casey B, Jacob OC. Vacuum catastrophe: An elementary exposition of the cosmological constant problem. American Journal of Physics. 1995;**63**:620-626

[38] Padmanabhan T. Cosmological constant: The weight of the vacuum. Physics Reports. 2003;**380**:235

[39] Hobson MP, Efstathiou GP, Lasenby AN. General Relativity: An Introduction for Physicists. New York, U.S.: Cambridge University Press; 2006

[40] Ryder LH. Quantum Field Theory. London, UK: Cambridge University Press; 1987

[41] Milonni PW. The Quantum Vacuum. New York: Academic Press Inc.; 1994

[42] Garisson JC, Chiao Y. Quantum Optics. New York: Oxford University Press; 2008

[43] Feynman RP. The Present Status of QED. Wiley Interscience: Hoboken, New Jersey; 1961

[44] Grynberg A, Aspect A, Fabre C. Quantum Optics. New York: Cambridge University Press; 2010

[45] Meis C. Light and Vacuum. 2nd ed. Singapore: World Scientific; 2017

[46] Sangwan A, Mukherjee A, Jassal HK. Reconstructing the dark energy potential. JCAM. 2018;**01**:018

[47] Caldwell R, Linder EV. The limits of quintessence. Physical Review Letters. 2005;**95**:141301

[48] Zlatev I, Wang L-M, Steinhardt PJ. Quintessence, cosmic coincidence, and the cosmological constant. Physical Review Letters. 1999;**82**:896-899

[49] Armendariz-Picon C, Mukhanov VF, Steinhardt PJ. A dynamical solution to the problem of a small cosmological constant and late time cosmic acceleration. Physical Review Letters. 2000;**85**: 4438-4441

[50] Garriga J, Mukhanov VF. Perturbations in k-ination. Physics Letters. 1999;**B458**:219-225

[51] Babichev E, Mukhanov V, Vikman A. k-Essence, superluminal propagation, causality and emergent geometry. JHEP. 2008;**0802**:101

[52] Wetterich C. The Cosmon model for an asymptotically vanishing time dependent cosmological constant. Astronomy & Astrophysics. 1995;**301**: 321-328

[53] Amendola L. Coupled quintessence. Physical Review D. 2000;**62**:043511

[54] Cai R-G, Wang A. Cosmology with interaction between phantom dark energy and dark matter and the coincidence problem. JCAP. 2005;**0503**: 002

[55] Copeland EJ, Liddle AR, Wands D. Exponential potentials and cosmological scaling solutions. Physical Review D. 1998;**57**:4686-4690

[56] Kamenshchik AY, Moschella U, Pasquier V. An alternative to quintessence. Physics Letters B. 2001; **511**:265-268

[57] Capozziello S, Carloni S, Troisi A. Quintessence without scalar fields.

Recent Research Developments in Astronomy and Astrophysics. 2003;**1**: 625. arXiv:astro-ph/0303041

[58] Bento M, Bertolami O, Sen A. Generalized Chaplygin gas, accelerated expansion and dark energy matter unification. Physical Review D. 2002; **66**:043507

[59] Song Y-S, Hu W, Sawicki I. Large scale structure of f(r) gravity. Physical Review D. 2007;**75**:44004

[60] Tsujikawa S. Matter density perturbations and effective gravitational constant in modified gravity models of dark energy. Physical Review D. 2007; **76**:023514

[61] Tsujikawa S. Observational signatures of f(R) dark energy models that satisfy cosmological and local gravity constraints. Physical Review D. 2008;**77**:023507

[62] Alnes H, Amarzguioui M, Gron O. An inhomogeneous alternative to dark energy? Physical Review D. 2006;**73**: 083519

[63] Iguchi H, Nakamura T, Nakao K-I. Is dark energy the only solution to the apparent acceleration of the present universe? Progress of Theoretical Physics. 2002;**108**:809-818

[64] Starobinsky AA. A new type of isotropic cosmological models without singularity. Physics Letters B. 1980;**91**: 99-102

[65] de la Cruz-Dombriz A, Dobado A. An f(R) gravity without cosmological constant. Physical Review D. 2006;**74**: 087501

[66] Nojiri S, Odintsov SD. Modifed f(R) gravity consistent with realistic cosmology: From matter dominated epoch to dark energy universe. Physical Review D. 2006;**74**:086005

[67] Nojiri S, Odintsov SD. Introduction to modified gravity and gravitational alternative for dark energy. International Journal of Geometric Methods in Modern Physics. 2006;**4**:06

[68] Rodrigues DC et al. Absence of fundamental acceleration scale in galaxies. Nature Astronomy. 2018;**2**: 668-672

[69] Boehmer CG, Harko T. Physics of dark energy particles. Foundations of Physics. 2008;**38**:216-227

[70] Wei H, Zou X-B, Li H-Y, Xue D-Z. Cosmological constant, fine structure constant and beyond. European Physical Journal C. 2017;**77**:14-27

[71] Cavalleri G, Barbero F, Bertazzi G, Cesaroni E, Tonni E, Bosi L, et al. A quantitative assessment of stochastic electrodynamics with spin (SEDS): Physical principles and novel applications. Frontiers of Physics in China. 2010;**5**(1):107-122

[72] Meis C. Quantized Field of Single Photons. 2019; DOI: 10.5772/intechopen.88378. Available from: https://www.intechopen.com/online-first/quantized-field-of-single-photons

[73] Lehmann H, Symanzik K, Zimermann W. Zur Formulierung quantisierter Feldtheorien. Nuovo Cimento. 1955;**1**:205-225

[74] Schwinger J, DeRaad LL Jr, Milton KA. Casimir effect in dielectrics. Annals of Physics. 1978;**115**:1

[75] Milonni PW. Casimir forces without the vacuum radiation-field. Physical Review A. 1982;**25**:1315

[76] Ezawa H, Nakamura K, Watanabe K. Frontiers in quantum physics. In: Lim SC, Abd-Shukor R, Kwek KH, editors. Proc. 2nd Int. Conference; 9–11 July 1997; Kuala

Lumpur, Malaysia. Springer; 1998. pp. 160-169

[77] Gründler G. The Casimir effect: No manifestation of zero-point energy. 2013. arXiv :1303.3790v5 [physics.gen-ph]

[78] Chuang SL. Physics of Photonic Devices. N.Y.: Wiley; 2009

[79] Read FH. Electromagnetic Radiation. N.Y.: Wiley; 1980

[80] Meis C. Vector potential quantization and the quantum vacuum. Physics Research International. 2014;ID 187432

[81] Meis C, Dahoo PR. Vector potential quantization and the photon wave-particle representation. Journal of Physics: Conference Series. 2016;**738**: 012099. DOI: 10.1088/1742-6596/738/1/012099

[82] Meis C, Dahoo PR. Vector potential quantization and the photon intrinsic electromagnetic properties: Towards nondestructive photon detection. International Journal of Quantum Information. 2017;**15**(8):1740003. DOI: 10.1142/S0219749917400032

[83] Meis C, Dahoo PR. Vector potential quantization and the photon wave function. Journal of Physics: Conference Series. 2017;**936**:012004. DOI: 10.1088/1742-6596/936/1/012004

[84] Meis C, Dahoo PR. The single-photon state, the quantum vacuum and the elementary electron/positron charge. American Institute of Physics Publications, Conference Proceedings. 2018;**2040**:020011. DOI: 10.1063/1.5079053

[85] Bialynicki-Birula I. Photon wave function. In: Wolf E, editor. Progress in Optics XXXVI. Amsterdam: Elsevier; 1996

[86] Chandrasekar N. Quantum mechanics of photons. Advanced Studies in Theoretical Physics. 2012;**6**: 391-397

[87] Sipe JE. Photon wave functions. Physical Review A. 1995;**52**:1875

[88] Smith BJ, Raymer MG. Photon wave functions, wave packet quantization of light and coherence theory. New Journal of Physics. 2007;**9**:414

[89] Khokhlov DL. Spatial and temporal wave functions of photons. Applied Physics Research. 2010;**2**(2):49-54

[90] Ehrenberg W, Siday RE. The refractive index in electron optics. Proceedings of the Physical Society. Section B. 1949;**62**:8-21

[91] Aharonov Y, Bohm D. Significance of electromagnetic potentials in the quantum theory. Physics Review. 1959;**115**:485-491

[92] Jackson JD. Classical Electrodynamics. N.Y.: Wiley; 1998

[93] Chambers RG. Shift of an electron interference pattern by enclosed magnetic flux. Physical Review Letters. 1960;**5**:3-5

[94] Tonomura A et al. Observation of Aharonov-Bohm effect by electron holography. Physical Review Letters. 1982;**48**(21):1443

[95] Osakabe N et al. Experimental confirmation of Aharonov-Bohm effect using a toroidal magnetic field confined by a superconductor. Physical Review A. 1986;**34**:815

[96] Bethe HA. The electromagnetic shift of energy levels. Physics Review. 1947; **72**(4):339-341

[97] Kroll NM, Lamb WE Jr. On the self-energy of bound electrons. Physics Review. 1949;**75**(3):388-396

[98] Milonni PW. Simplified derivation of the Hawking-Unruh temperature for

an accelerated observer in vacuum. American Journal of Physics. 2004; **72**(12):1524-1529

[99] Fulling SA. Nonuniqueness of canonical field quantization in Riemannian space-time. Physical Review D. 1973;7(10):2850-2862. DOI: 10.1103/PhysRevD.7.2850

[100] Davies PCW. Scalar production in Schwarzschild and Rindler metrics. Journal of Physics A. 1975;**8**(4):609-616. DOI: 10.1088/0305-4470/8/4/022

[101] Unruh WG. Notes on black-hole evaporation. Physical Review D. 1976; **14**(4):870-892. DOI: 10.1103/PhysRevD.14.870

[102] Visser M. Experimental Unruh radiation? Matters of Gravity. 2001;**17**: 4-5

[103] Ford GW, O'Connell RF. Is there Unruh radiation? Physics Letters A. 2005;**350**(1–2):17-26

[104] Greulich KO. Calculation of masses of all fundamental elementary particles with an accuracy of approx. 1%. Journal of Modern Physics. 2010;**1**:300-302

[105] Lieu R, Hillman LW. Probing Planck scale physics with extragalactic sources? Astrophysical Journal Letters. 2003;**585**:L77

[106] Götz D, Laurent P, Antier S, Covino S, D'Avanzo P, D'Elia V, et al. GRB 140206A: The most distant polarized gamma-ray burst. Monthly Notices of the Royal Astronomy Society. 2014;**444**:2776

[107] Meis C. The electromagnetic field ground state and the cosmological evolution. Journal of Physics, Conference Series. 2018;**1141**:012072. DOI:10.1088/1742-6596/1141/1/012072

Chapter 6

Dark Matter as Cold Atomic Hydrogen in Its Lower Ground State

Eugene Terry Tatum

Abstract

This chapter presents a novel theory of dark matter made plausible by several astronomical observations reported in 2018 and 2019. The author introduced this theory to colleagues invited to the Dark Matter Workshop at the World Science Festival in May of 2019, and its first publication was in a peer-reviewed physics journal in July of 2019.

Keywords: dark matter, atomic hydrogen, interstellar medium, cosmic dawn, Wouthuysen-Field effect, cosmology theory, Milky Way galaxy

1. Introduction

The theory [1], simply stated, is that what we currently refer to as "cold dark matter" is, in actuality, slow-moving interstellar and intergalactic neutral atomic hydrogen in its lower 1 s ground state. Its exceedingly low density within the vacuum of space can be quantified by measuring the intensity of its signature spectral hyperfine 21-cm *absorption* line in lines of site to stellar objects at known distances. At an average HI density of approximately one atom per cubic centimeter (1.67×10^{-21} kg m^{-3}) within the vast, cold, and remote interstellar vacuum of the Milky Way (MW), it is very nearly collisionless and thus mostly unperturbed. And, given its current nearly perpetual lower ground state condition, it *cannot* emit light. Whenever and wherever hydrogen is mostly above this ground state, and significantly more concentrated, it is readily visible and we call it something else (a cold, warm, or hot gas cloud, for instance).

Following a brief review of the historical evidence for the existence of dark matter, its key observations reported in 2018 and 2019 will be summarized and its current constraints elaborated. The author's calculations, in the context of these observations, will then be presented in the Results section, and a Discussion section with a table based upon these findings will follow.

2. Historical background

It is generally agreed that astronomer Fritz Zwicky, in 1933, was the first scientist to apply the virial theorem to infer the existence of dark matter. He referred to it as "dunkle materie" (i.e., "dark matter") [2, 3]. Unfortunately, Zwicky's dark matter proposal was largely ignored at that time.

Beginning in 1970, this problem of "missing matter" was further elucidated and essentially proven by the detailed studies of galactic rotation by Vera Rubin and William Ford [4, 5], although it took considerable time for them to receive due recognition for this achievement.

With gradual acceptance of the observational implications, what has followed in the ensuing decades has been a stepwise progression of tightening constraints on the nature and quantity of dark matter. As a consequence, much like a horse race with changing leads, various creative and exotic theories of the nature of dark matter (WIMPs, MACHOs, axions, sterile neutrinos, supersymmetry partners, SIMPs, GIMPs, etc.) have fallen in and out of favor [6]. Given the difficulty of its detection, there have even been attempts to discard the idea of dark matter altogether in favor of modifying Newtonian celestial mechanics (Modified Newtonian Dynamics (MOND)) [7, 8].

A review of these various theories, and a discussion of their current plausibility, is beyond the scope of this chapter. Whole books have been written about them. Suffice it to say, in view of the many continuing exotic dark matter detector failures, there is room for a new theory such as the one presented herein. The following section will summarize key constraints on dark matter as of 2020.

3. Current observational constraints

Upon establishing the likelihood of an abundance of cosmic matter which, *in its current state*, does not emit light, astronomers and astrophysicists have attempted to quantify it with respect to the visible matter (i.e., stars, gas clouds, and cosmic dust). The 2018 Planck Collaboration report [9] indicates a cosmic dark matter-to-visible matter ratio of approximately 5.4:1. This is in close agreement with a ratio of approximately 5:1 established by a 2019 *Gaia*-Hubble survey report [10] on the Milky Way galaxy. The *Gaia* report indicates a total virial MW mass of approximately 1.5 trillion solar masses which include a visible matter mass of approximately 250 billion solar masses. Based upon these and other studies, dark matter is currently believed to comprise about 85% of all cosmic matter. Thus, although it appears, by gravitational lensing, to be predominantly within and haloed around the visible galaxies, dark matter is most likely ubiquitous and therefore a key structural (i.e., "scaffold") component of the universe. In this context, it is worth noting that the Planck Collaboration study of the cosmic microwave background (CMB) anisotropy documents the presence and gravitational influence of dark matter within the hot and dense early universe during the recombination/decoupling epoch. So what we now tend to think of as "cold dark matter" (CDM) was once hot, and very possibly light-emitting, in its past excited state.

Although relatively few in number, MW halo stars at various known distances beyond the galactic disk can provide for line-of-site spectral analysis and a rough MW halo vacuum density determination of interstellar neutral atomic hydrogen in its lower ground state. Specifically, the intensity of the hyperfine 21-cm absorption line gives us some idea of the number of these particular atoms per unit volume of the column of intervening interstellar space. Best estimates of this sort, made over a number of decades, have indicated an average density within the MW interstellar vacuum of roughly one of these atoms per cubic centimeter [11–13].

Making use of some initial *Gaia* survey data released in 2018, Posti and Helmi reported results [14] which allow one to deduce a ratio of dark matter-to-visible matter within a 20 kpc (i.e., 65,000 light-years) radius halo sphere of the MW (see schematic **Figure 1**). This halo sphere is represented in black in the figure and is

Figure 1.
Posti and Helmi 20 kpc halo sphere of the MW galaxy.

roughly to scale with respect to the 50,000 light-year radius MW disk (in white). The disk averages approximately 1000 light-years in thickness. The relatively few halo stars well beyond the disk are also schematically represented in the figure. As mentioned, these are useful for density measurements of cold hydrogen in the lower ground state within the halo vacuum.

The total virial mass of their sphere was reported by Posti and Helmi to be 1.91×10^{11} $M_\odot$ (solar masses), of which the mass of dark matter was reported to be 1.37×10^{11} $M_\odot$. This would imply that the MW 20 kpc sphere ratio of dark matter-to-visible matter is about 2.54:1. Therefore, if we normalize the MW visible mass to the 250 billion $M_\odot$ value given in the 2019 *Gaia* survey report, this Posti and Helmi ratio would imply a corresponding dark matter mass of approximately 635 billion $M_\odot$ within the same 20 kpc radius halo sphere. These numbers will be compared in the subsequent Results section.

Aside from the inability of dark matter to emit light, observations have confirmed that it is nearly collisionless. It appears to be composed of particles with a low scattering cross section. This can be deduced from Tucker's early observations of the bullet cluster [15] and subsequent observations of other colliding galaxies.

Dark matter, at present, is also believed to be cold (i.e., slow-moving). A predicted Maxwell-Boltzmann particle velocity distribution ranging from roughly 0 to 600 km/sec, and peaking at roughly 220–230 km/sec, is the theoretical basis for optimizing a variety of cold dark matter particle detectors [16]. Unfortunately, none of these experiments to date has produced a positive result of an exotic (i.e., non-baryonic) dark matter particle. Intriguingly, however, the 2018 EDGES study [17] of the hyperfine 21-cm *absorption* line of neutral atomic hydrogen corresponding to cosmological redshifts of $15 < z < 20$ (cosmic dawn) has reported a strong signal consistent with a hydrogen gas temperature in the low single digits of the Kelvin temperature scale. This is considerably lower than the cosmic dawn CMB radiation temperature and produces strong constraints on the nature of dark matter. This *CMB decoupling phase* during cosmic dawn indicates that whatever we are currently referring to as dark matter has been particularly cold since at least the time of early cosmic dawn, has a particle mass of no more than about 2–3 GeV, and has a scattering cross-sectional σ_1 value of at least 1.5×10^{-21} cm^2. If the EDGES observations of cosmic dawn are, in fact, the result of dark matter cooling of warmer (i.e., CMB-equilibrated) hydrogen atoms, the proposed WIMPs and all but one baryon (namely, *colder* atomic hydrogen in its lower ground state) are effectively ruled out as dark matter candidates.

Figure 3 on page 9 of Barkana's review [18] related to the EDGES study findings summarizes the new cosmic dawn dark matter constraints with a log graph of the implied baryon-dark matter (b-DM) cross-sectional σ_1 and the minimum possible 21-cm brightness temperature (T_{21}) on the two vertical axes and the corresponding implied dark matter particle mass M_X on the horizontal axis. All constraint values indicated in the graph correspond to the strong signal measured at $z = 17$, which corresponds to a redshifted 21-cm hyperfine hydrogen absorption line detectable at a frequency of 78.9 MHz. To fully comprehend the significance of these dark matter constraints, the reader should obtain this reference and pay particular attention to the dark matter particle mass corresponding to a cross-sectional σ_1 value of 10^{-20} cm^2 and a 21-cm brightness temperature $\log_{10}$ value (in mK) of 2.32. Please note that *these values correspond to a cold dark matter particle fitting with neutral atomic hydrogen, which has a similar low velocity scattering cross section and a mass energy of 0.938 GeV. Furthermore, it should be remembered that the 21-cm absorption line is the signature of atomic hydrogen in its lower ground state. These new cosmic dawn constraints on dark matter will be a major focus in the following Discussion section, particularly with respect to the Wouthuysen-Field effect.*

Without specifically naming any particular non-excluded baryons, physicist Stacy McGaugh published a brief note [19] at the time of the EDGES publication (March 2018) which strongly supports the idea that the cosmic dawn observations are to be, in his words, "expected for a purely baryonic universe." He begins the note with the observation that the strength of the redshifted hyperfine 21-cm absorption line at $z = 17$ is anomalously strong for ΛCDM, which proposes non-baryonic dark matter. He also points out that current knowledge in atomic physics would indicate that a maximum possible T_{21} signal should occur when the neutral hydrogen fraction $X_{HI} = 1$ and spin temperature $T_S = T_K$. McGaugh's cogent arguments and interpretation of the EDGES cosmic dawn data are strongly supportive of the theory presented herein.

An additional constraint on dark matter has to do with the "cusp-core problem," specifically why some galaxies have a distinctly cuspy distribution of dark matter and others do not. A 2019 report on dark matter distribution within dwarf galaxies, by Read et al. [20], offers a clue. It shows that galaxies which stopped forming stars over 6 billion years ago tend to be cuspier than those with more extended star formation. This is equivalent to saying that the extended star formation dwarf galaxies have shallower dark matter cores. Thus, their findings agree well with models where dark matter is presumably heated up by bursty star formation. This means that any plausible theory of dark matter must explain why extended and bursty star formation is correlated with a so-called "cored" dark matter distribution.

One obvious possible interpretation of the Read observations is simply that bursts of highly energetic particles and photons, produced by a concentration of new stars in and around active galactic centers, would tend to heat up and eject cold dark matter from their vicinity. If this is the correct interpretation, then a self-interacting dark matter (SIDM) model becomes unnecessary to explain the "cusp-core problem." In fact, all sorts of bizarre non-baryonic properties of dark matter then become unnecessary.

4. Results (calculation)

Given the new dark matter theory as briefly summarized in the Introduction section, a simple calculation can be made on the Posti and Helmi 20 kpc MW halo sphere, as a test of this theory. If we start with the current best estimate of an average of only one atom of atomic hydrogen in the lower ground state per cubic centimeter of the Posti and Helmi 20 kpc halo sphere, that assumes a vacuum hydrogen density of 1.67×10^{-21} kg m^{-3}. If we then multiply that number by the volume of the 20 kpc sphere (9.85×10^{62} m^3), the total mass of atomic hydrogen in the bottom ground state is 1.645×10^{42} kg. That is the equivalent of 827 billion $M_\odot$. This is 3.3 times the 2019 *Gaia* survey MW galaxy visible mass! Even allowing for only 0.75 atom of atomic hydrogen in the bottom ground state per cubic centimeter of the 20 kpc halo sphere, the Posti and Helmi dark matter-to-visible matter ratio of 2.54 can be met.

5. Discussion: interstitial hydrogen, cosmic dawn, and the Wouthuysen-Field effect

Observations of the CMB anisotropy map suggest the following cosmic evolution scenario since the CMB emission epoch:

Denser regions of the primordial hydrogen distribution, already subject to the positive feedback of gravity, further aggregated into the hot stars, warm gas clouds, galaxies, quasars, and filaments. In contrast, due to adiabatic cosmic expansion, the primordial hydrogen within the low gravity interstices of the CMP map progressively became exceedingly sparse and cold (i.e., CMB-equilibrated). These interstices we know today as the vast interstellar and intergalactic space, including the voids.

The expanding and cooling universe, after CMB emission, was completely dark before the first dense clusters of primordial hydrogen underwent nuclear fusion. This period, known as the cosmic "dark age," merged into the "cosmic dawn" reionization epoch at around 100 million years after the big bang. The "cosmic dawn" epoch is named as such because this is when the first stars are thought to have formed.

As documented by the EDGES study, a strange phenomenon occurred during the period of cosmic dawn. For about 150 million years, corresponding roughly to the cosmological redshift range of $15 < z < 20$, the temperature T_G of the vast interstitial primordial hydrogen gas was *decoupled* from the CMB radiation temperature T_R. At the peak of this phenomenon, at roughly $z = 17$, this primordial hydrogen appears to have been in the low single digits of the Kelvin temperature scale. Thereafter, the hydrogen gas gradually warmed back up to the CMB temperature at roughly $z = 15$. **Figure 2** illustrates this phenomenon. On this graph, $z = 20$ corresponds to about 100 million years after the big bang, $z = 17$ corresponds to about 180 million years after the big bang, and $z = 15$ corresponds to about 250 million years after the big bang.

This phenomenon of "cosmic dawn CMB decoupling" is most commonly attributed to a b-DM scattering interaction, whereby dark matter is presumed to have chilled faster than primordial hydrogen during the cosmic dark age, to the point where it could then interact with and chill the CMB-equilibrated interstitial hydrogen and decouple it from the CMB radiation temperature.

The problem with this particular explanation of the EDGES study observations is to explain why the beginning of the CMB decoupling phenomenon *coincided with the first stars* at the crack of cosmic dawn. How is it that dark matter had cooled sufficiently to enable b-DM scattering and CMB decoupling of primordial hydrogen

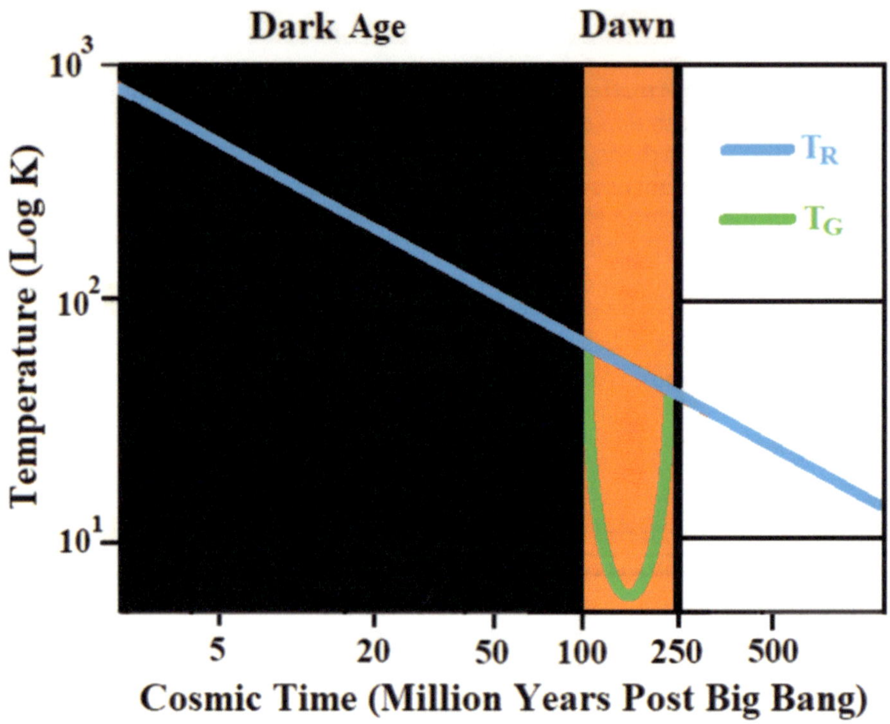

Figure 2.
Cosmic dawn CMB decoupling of primordial hydrogen.

just when the first stars were forming? Could there be a simpler explanation for cosmic dawn CMB decoupling *without requiring a non-baryonic intermediary?*

Fully in keeping with McGaugh's bold assertion of a purely baryonic mechanism, this cosmic dawn coincidence may have been *entirely* due to the Wouthuysen-Field (WF) effect on CMB-equilibrated primordial atomic hydrogen. If unfamiliar with this radiation effect on atomic hydrogen, the reader is encouraged to read an excellent and brief summary of the WF effect on the Wikipedia page entitled "Wouthuysen-Field Coupling" [21]. A more extensive and highly technical summary is also found on the AstroBaki website [22]. Briefly, the Lyman-alpha ultraviolet (UV) radiation of the first stars was of sufficient energy to have caused a redistribution of the balance of the two hydrogen electron hyperfine 21-cm ground states such that the primordial hydrogen gas could effectively bypass its "forbidden transition" (from parallel to antiparallel electron spin) and *easily reach the lower ground state.* The net effect of this process would have been to decouple primordial hydrogen from the CMB radiation temperature, producing the strong 21-cm absorption line signal observed. Thus, it appears that *an exotic, non-baryonic, form of dark matter was completely unnecessary for cosmic dawn CMB decoupling. The mysterious dark matter at cosmic dawn could simply have been the first of the interstitial hydrogen to be chilled and decoupled by the Lyman-alpha radiation. The process then, over millions of years, would have extended to the rest of the CMB-equilibrated hydrogen, peaking at a cosmic redshift of $z = 17$.*

The key dark matter features, including observational constraints achieved over the last few years, and the correlating features of interstitial atomic hydrogen in the lower HI ground state, can now be brought together into a table (**Table 1**) for comparison.

These correlations are striking and *strongly suggest that interstitial cold atomic hydrogen in its lower ground state is what we have been calling dark matter over the last few decades.*

Dark matter features	Interstitial HI cold hydrogen	References
Cold (0–600 km/sec)	Cold (0–600 km/sec)	[16]
Dark (no emissions)	Lower ground state (cannot emit)	[2–5]
Cross-section $\sigma_1 > 1.5 \times 10^{-21}$ cm	$\sigma_1 > 1.5 \times 10^{-21}$ cm (at low velocity)	[15, 17, 18]
Baryon (strongest 21-cm signal)	Baryon for $X_{HI} = 1$ and $T_S = T_K$	[19]
Mass-Energy less than 3 GeV	Mass-Energy = 0.938 GeV	[17, 18]
Mass 20 kpc Halo = 635 Billion $M_\odot$	Mass 20 kpc Halo = 827 Billion $M_\odot$	[14, 10]
Central DM heating ("coring")	Ejected and loses ground state	[20]
CMB decoupling at cosmic dawn	Wouthuysen-Field Effect	[21, 22]
Structural scaffold	Most abundant atom	[9]
Existence at CMB emission	Most abundant atom	[9]

Table 1.
Dark matter features vs. interstitial HI cold hydrogen.

It has long been assumed that the average atomic density of the "nearly empty" vacuum of interstellar space beyond the visible stars, gas clouds, and cosmic dust can be ignored in galactic mass calculations. While this might be true for the confines of the galactic disk and bulge, where visible matter is particularly concentrated, it is definitely not true for the galactic halo in close proximity to the disk. The sheer vastness of space belies the assumption mentioned above. It appears that this mistaken assumption has been a key foundational error behind the long-standing mystery of dark matter. The simple calculation in the Results section supports this conclusion. Even a single stray baryonic atom per cubic centimeter of interstellar space within the 20 kpc MW halo of Posti and Helmi can dwarf the combined mass of all visible stars, clouds of gas, and cosmic dust!

The fact that the particular atom in question appears now not to be in the least bit exotic but, instead, the most common structural element in the universe is indeed ironic. In a sense, because of the many distractions and obscurations provided by the highly visible warm and hot hydrogen atoms, cold interstitial hydrogen, because of its remote location, extremely low density, low velocity, and prolonged lower ground state, has been *essentially hiding in plain sight.* Observations of the 21-cm hyperfine *absorption* line (its *signature*) have been noted for decades but only recently connected to phenomena attributed to dark matter.

Any useful physical theory should be falsifiable and predictive. The falsifiability of this particular theory is obvious. This theory would be falsified if a particle M_X of 0.938 GeV becomes excluded from dark matter constraints, or current best estimates of the average MW halo vacuum density of cold atomic hydrogen are subsequently proven to be *severely* overestimated. However, a minor correction to approximately 0.5–0.75 atom per cubic centimeter is entirely consistent with this theory. As for observations to further strengthen this theory, the following predictions are made:

1. There will be tightening dark matter constraints around a particle M_X value of 0.938 GeV (i.e., the mass energy of neutral atomic hydrogen).

2. Computer simulations of galaxy formation and evolution which incorporate this theory will show excellent correlations with observations, including the coring effect of heating and ejecting cold interstellar hydrogen from active galactic centers with bursty star formation.

3. No exotic non-baryonic particles fitting the observed qualitative and quantitative constraints will ever be discovered.

6. Summary and conclusion

To summarize, this chapter has introduced the reader to a plausible new theory of dark matter which appears to match current observational constraints. The theory, simply stated, is that what we currently refer to as "cold dark matter" is, in actuality, slow-moving interstellar and intergalactic neutral atomic hydrogen in its lower 1 s ground state. So long as it stays in this lower ground state, it cannot emit light. Furthermore, it is currently so sparse as to be nearly collisionless. Whenever and wherever hydrogen is mostly above this ground state, and significantly more concentrated, it is readily visible and we call it something else (a cold, warm, or hot gas cloud, for instance).

Dark matter observations corresponding to the cosmic dawn epoch, which were reported in 2018 and 2019, have provided the necessary constraints on dark matter to favor this theory above all others at the present time. In particular, the Bowman (i.e., EDGES) and Barkana references point to a cold dark matter particle with features quite consistent with cold atomic hydrogen. Furthermore, a convincing case has been made by McGaugh that the strong hydrogen absorption signal at cosmic dawn is the *signature* of a baryonic universe. The obvious mechanism for such signal strength, and its coincidence with cosmic dawn, is the Wouthuysen-Field effect. From the forgoing discussion, it becomes apparent that exotic (i.e., non-baryonic) matter is not necessary to explain dark matter observations to date.

We conclude by asking the following question:

If interstitial cold atomic hydrogen in its lower ground state is qualitatively and quantitatively sufficient to explain dark matter observations to date, do we really need to spend more of our time and money continuing to look for anything else?

Author details

Eugene Terry Tatum
Independent Researcher, Bowling Green, KY, USA

*Address all correspondence to: ett@twc.com

References

[1] Tatum ET. My C.G.S.I.S.A.H. Theory of dark matter. Journal of Modern Physics. 2019;**10**:881-887. DOI: 10.4236/jmp.2019.108058

[2] Zwicky F. Die Rotverschiebung von extragalaktischen Nebeln. Helvetica Physica Acta. 1933;**6**:110-127. Bibcode: 1933AcHPh...6..110Z

[3] Zwicky F. On the Masses of Nebulae and of Clusters of Nebulae. Astrophysical Journal. 1937;**86**:217. Bibcode: 1937ApJ....86..217Z. DOI: 10.1086/143864

[4] Rubin V, Ford WK. Rotation of the Andromeda Nebula from a spectroscopic survey of emission regions. The Astrophysical Journal. 1970;**159**:379. DOI: 10.1086/150317

[5] Rubin V et al. Rotation velocities of 16 SA galaxies and a comparison of Sa, Sb, and SC rotation properties. The Astrophysical Journal. 1985;**289**:81. DOI: 10.1086/162866

[6] Hadhazy A. The Dark Matter Derby. Discover. 2019;**40**(9):40-47

[7] Milgrom M. A modification of the Newtonian dynamics as a possible alternative to the hidden mass hypothesis. The Astrophysical Journal. 1983;**270**:365-370. Bibcode: 1983ApJ...270..365M. DOI: 10.1086/161130

[8] Kroupa P, Pawlowski M, Milgrom M. The failures of the standard model of cosmology require a new paradigm. International Journal of Modern Physics D. 2012;**21**(14):1230003. arXiv:1301.3907. Bibcode: 2012IJMPD..2130003K. DOI: 10.1142/S0218271812300030

[9] Aghanim N et al. Planck 2018 Results VI. Cosmological Parameters; 2018. Available from: http://arXiv:1807.06209v1

[10] Watkins LL, et al. Evidence for an intermediate-mass Milky Way from *Gaia* DR2 Halo Globular Cluster Motions. The Astrophysical Journal. 2019;**873**:118-130. DOI: 10.3847/1538-4357/ab089f

[11] Mammana DL. Interstellar Space. New York: Popular Science; 2000. p. 220

[12] Chaisson E, McMillan S. Astronomy Today. New York: Prentice Hall; 1993. p. 418

[13] Pananides NA, Arny T. Introduction to Astronomy. 2nd ed. New York: Wiley & Sons; 1979. p. 293

[14] Posti L, Helmi A. Mass and shape of the Milky Way's dark matter Halo with globular clusters from *Gaia* and Hubble. Astronomy & Astrophysics. 2019;**621**:A56. DOI: 10.1051/0004-6361/201833355

[15] Tucker W. Recent and Future Observations in the X-ray and Gamma-ray Bands: Chandra, Suzaku, GLAST, and NuSTAR. AIP Conference Proceedings. 2006;**801**(21). arXiv:astro-ph/0512012. DOI: 10.1063/1.2141828

[16] Baushev AN. Principal properties of the velocity distribution of dark matter particles near the solar system. Journal of Physics Conference Series. 2012;**375**:012048. DOI: 10.1088/1742-6596/375/1/012048

[17] Bowman JD. An Absorption profile centered at 78 megahertz in the sky-averaged spectrum. Nature. 2018;**555**:67. DOI: 10.1038/nature25792

[18] Barkana R. Possible interactions between baryons and dark matter. 2018. arXiv:1803.06698v1 [astro-ph.CO]

[19] McGaugh SS. Strong Hydrogen Absorption at Cosmic Dawn: The Signature of a Baryonic Universe. 2018. arXiv:1803.02365v1 [astro-ph.CO]. DOI: 10.3847/2515-5172/aab497

[20] Read JI, Walker MG, Steger P. Dark Matter Heats Up in Dwarf Galaxies. Monthly Notices of the Royal Astronomical Society. 2019;**484**:1401-1420. arXiv:1808.06634v2 [astro-ph.CO]. DOI: 10.1093/mnras/sty3404

[21] Wikipedia Contributors. Wouthuysen-Field Coupling. Wikipedia, The Free Encyclopedia; 2019. Available from: https://en.wikipedia.org/w/index.php?title=Wouthuysen%E2%80%93Field_coupling&oldid=913891210 [Accessed: 04 February 2020]

[22] AstroBaki. 2017. Available from: https://casper.ssl.berkeley.edu/astrobaki/index.php/Wouthuysen_Field_effect